FRONTIERS IN RADIATION ONCOLOGY

Edited by **Tejinder Kataria**

Frontiers in Radiation Oncology

http://dx.doi.org/10.5772/3065
Edited by Tejinder Kataria

Contributors

Takahiro Oike, Hideaki Ogiwara, Takashi Nakano, Takashi Kohno, Jin-Peng Qi, Yong-Sheng Ding, Xian-Hui Zeng, Thatiane Alves Pianoschi, Mirko Salomón Alva-Sánchez, Mirko Salomón Alva-Sánchez, Thatiane Alves Pianoschi, Kei Ichiji, Noriyasu Homma, Masao Sakai, Makoto Abe, Norihiro Sugita, Makoto Yoshizawa, S. Agustín Martínez Ovalle, Chul-Seung Kay, Young-Nam Kang, Hend Ahmed El-Hadaad, Hanan Ahmed Wahba, Simonetta Franchelli, Paolo Meszaros, Michela Massa, Marina Guenzi, Renzo Corvò, Davide Pertile, Giorgia Timon, Ferdinando Cafiero, Pierluigi Santi, Jimmy Yu-Wai Chan, Gregory Ian Siu Kee Lau, Sabina Saric, Elisabeth Isenring

Published by InTech

Janeza Trdine 9, 51000 Rijeka, Croatia

Publishing Process Manager Oliver Kurelic

Typesetting InTech Prepress, Novi Sad

Cover InTech Design Team

Additional hard copies can be obtained from orders@intechopen.com

Frontiers in Radiation Oncology, Edited by Tejinder Kataria
p. cm.
ISBN 978-953-51-1163-4

For EU product safety concerns:
IN TECH d.o.o., Prolaz Marije Krucifikse Kozulić 3, 51000 Rijeka, Croatia,
info@intechopen.com or visit our website at intechopen.com.

Meet the editor

Dr Tejinder Kataria is the Chairperson, Division of Radiation Oncology at Medanta-the Medicity, Gurgaon, India. After her graduation she pursued MD in Radiotherapy followed by a fellowship for intraluminal brachytherapy. She then went on to hone her skills in Brachytherapy for flexible Iridium 192 interstitial implants and permanent Iodine125 seed implants for prostate, while on a Commonwealth scholarship. She returned to India and started the department of Radiotherapy in one of the premier cancer centers in Delhi. She trained at DKFZ and implemented Intensity Modulated Radiotherapy in the country in 2001-02.Subsequently the quest for learning took her to Mannheim Klinikum, in 2007 for image guided radiotherapy and in 2010 for volumetric arc therapy . She started frameless Linac based stereotaxy in 2010 after training at Wurzburg .She was instrumental in bringing the first the Cyberknife- VSI system to her country. Her efforts have been towards implementing conformal radiation therapy techniques and advancements for the benefit of her patients and training the younger generation in the art and science of cancer care.

Contents

Preface

Radiation therapy was applied for treatment of cancers empirically after the discovery of radium . The acute and long term side effects were recorded clinically as the patients started outliving the tumours. The combination of radiation before or after surgery and subsequent addition of chemotherapy has revolutionized the way we treat cancer today. Every tumour is approached through a multimodality discussion such that the toxicity of either treatment can be reduced with maximum curative potential. **Frontiers in Radiation Oncology** has been brought forth to understand the basics of radiation sensitization, cellular and genomic stress responses to radiation,inhibiting repair of subletahl damage along with an understanding of the dosimetric aspects of radiation physics. The chapters on clinical aspects have been designed to bring out the changing concepts of cure in metastatic disease with the advent of stereotactic body radiotherapy. The effects of radiation, concurrent chemotherapy and surgery are also the subjects of exploration in the clinical context. Any textbook on providing care to cancer patients cannot be complete without addressing the quality of life in cancer treatment and the nutritional needs of the high catabolic state in the cancer patients. The last two chapters address these needs.

The book is meant for physicians, physicists, dosimetrists and counselors who are trying to provide a holistic care in cancer while trying to understand the complex basics of radiation interaction within the human body.

Tejinder Kataria
Radiation Oncology, Medanta-The Medicity,
India

Section 1

Molecular Biology of Radiation Therapy

Chapter 1

Histone Acetyltransferases (HATs) Involved in Non-Homologous End Joining as a Target for Radiosensitization

Takahiro Oike, Hideaki Ogiwara, Takashi Nakano and Takashi Kohno

Additional information is available at the end of the chapter

http://dx.doi.org/10.5772/56555

1. Introduction

Radiation therapy is one of the most important treatment modalities for cancer therapy alongside surgery and chemotherapy. However, a major problem associated with radiation therapy in the clinical setting is that, in many cases, local control of the tumor cannot be achieved using this modality alone. This has driven researchers into radiosensitizers, i.e., compounds that enhance the intrinsic sensitivity of cancer cells to ionizing radiation (IR). Several compounds, including halogenated pyrimidines and nitroimidazole derivatives, show radiosensitizing effects in cancer cells [1]; however, clinical application of these compounds is limited because they are highly toxic to normal cell and tissues. Therefore, radiosensitizers with low toxicity to normal tissues are urgently needed.

The principal target for IR-induced killing of cancer cells is DNA [2]. Of the different types of DNA damage generated by IR, DNA double-strand breaks (DSBs) are thought to be the most cytotoxic. DSBs induced by IR are preferentially repaired by non-homologous end joining (NHEJ) [3, 4], which joins the two broken DNA ends without the need for sequence homology. To enable NHEJ, the chromatin needs to be remodeled into an 'open' state so that the DNA repair proteins can access the DSB sites [5]. We previously reported that acetylation of histone proteins at DSB sites by the histone acetyltransferases (HATs), TIP60, CBP and p300, facilitates NHEJ through chromatin remodeling [6]. This suggests that the inhibition of HAT activity will radiosensitize cancer cells by suppressing NHEJ. In line with this, we and others demonstrated that several natural compounds with HAT-inhibitory activity are able to radiosensitize cancer cells [6-12]. Since some of these compounds are safe when administered to humans [13-15], they could potentially be used as radiosensitizers in a clinical setting. Here, we discuss the role of HATs in NHEJ, the radiosensitizing effects of compounds with HAT-inhibitory activity, and the prospects for the clinical application of these compounds.

2. Radiosensitization by HAT inhibition

2.1. HATs are involved in chromatin remodeling required for DNA repair

Chromosomal DNA and histones form a highly condensed structure known as chromatin. During the repair of DNA DSBs, the accessibility of DNA repair proteins to the DSB sites on chromosomal DNA is regulated by the relaxation of the chromosome structure via chromatin remodeling. Remodeling is mediated by both covalent (histone modifications, e.g., acetylation) and non-covalent (ATPase-dependent chromatin remodeling) interactions **(Figure 1)**. Several studies show that the acetylation of histones located at the DSB sites is a critical step for DNA repair [16-18]; however, the HATs involved in NHEJ have not been fully identified.

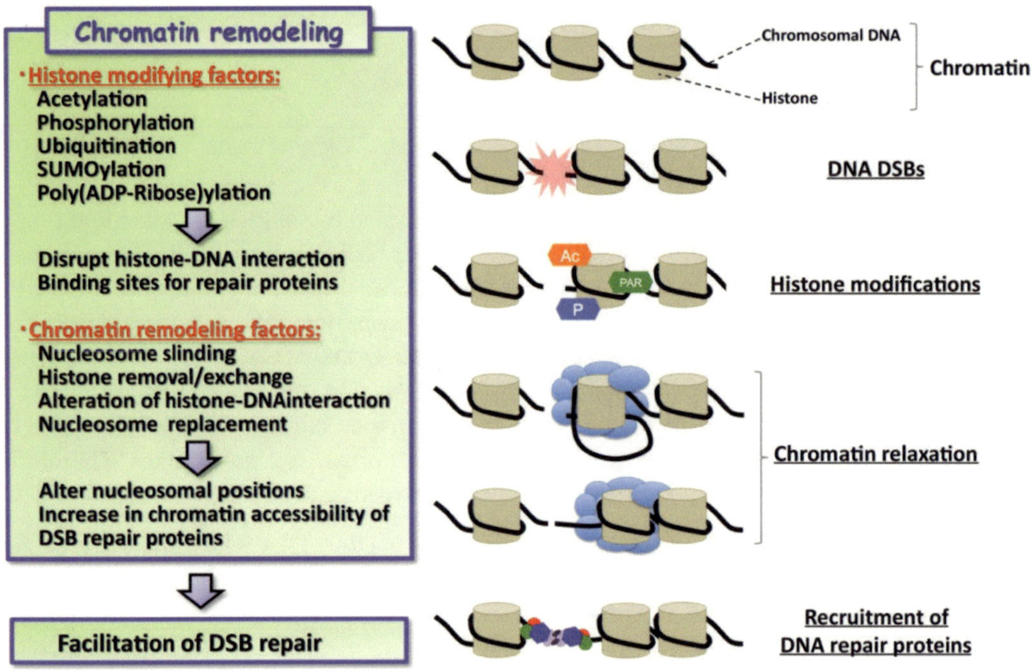

Figure 1. Chromatin remodeling is required for DSB repair. Both histone modifications and ATPase-dependent chromatin remodeling are needed for efficient repair.

2.2. Cell-based NHEJ activity assay

We developed a new assay system for evaluating NHEJ repair of DSBs in the chromosomal DNA in living human cells [6, 18] because the existing NHEJ assays used only non-chromosomal (i.e., plasmid) DNA. The assay design is outlined in **Figure 2**. The IRES-TK-EGFP plasmid, which contains two recognition sites for *I-Sce*I endonuclease [20] in the reverse direction, was integrated into the chromosomal DNA of H1299 human lung cancer cells as a substrate for DSBs and subsequent NHEJ repair. Human genomic DNA does not contain *I-Sce*I sites; therefore, the *I-Sce*I protein transiently expressed after transfection of the *I-Sce*I expression plasmid specifically cleaves the two *I-Sce*I sites in the substrate DNA to yield DSBs with incompatible ends. This results in DSBs in the chromosomal DNA. NHEJ of the two broken DNA strands results in deletion of the herpes simplex virus-thymidine kinase (TK) open reading frame and leads to the production of a transcript that enables the translation of enhanced green fluorescent protein (EGFP) instead of the TK protein.

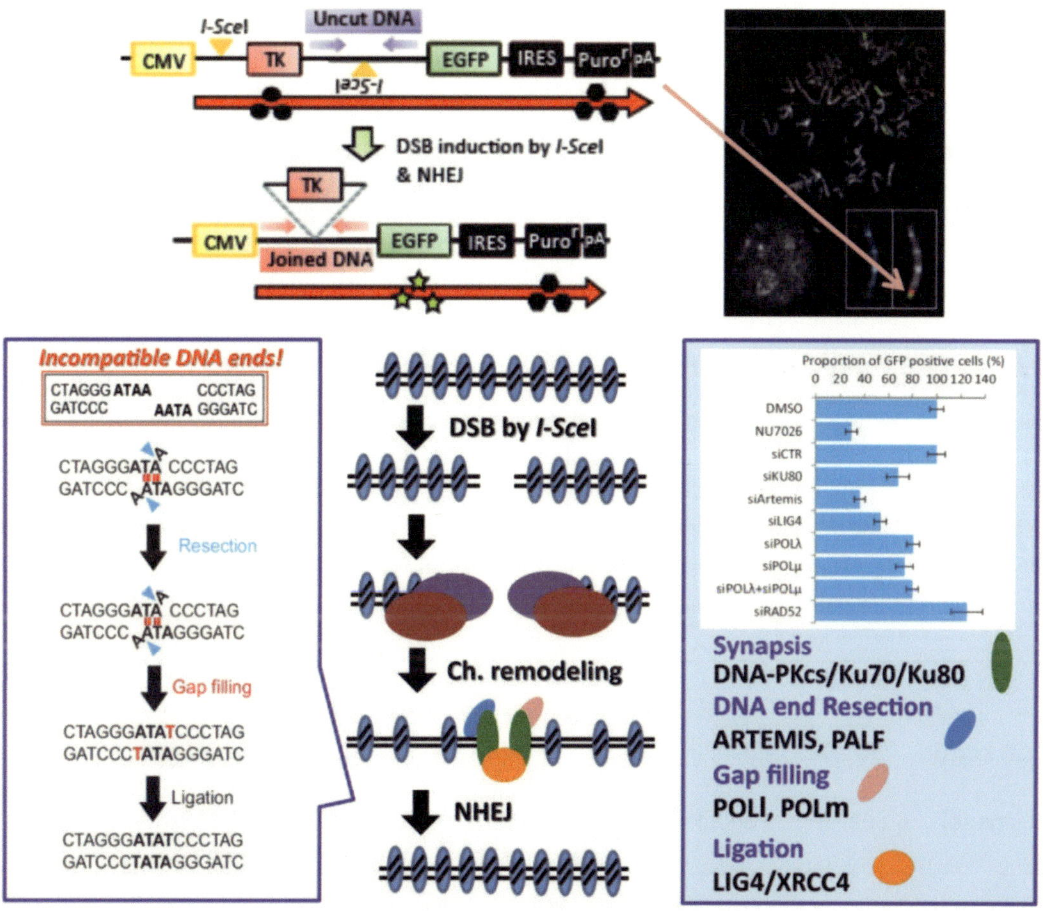

Figure 2. Cell-based NHEJ activity assay [6]. Upper panel: Assay design. Two *I-Sce*I sites (in the reverse direction) are indicated by the yellow arrow heads. The locations of the PCR primers used for quantitative PCR to monitor DSB introduction by *I-Sce*I (uncut DNA) and subsequent joining (joined DNA) are indicated by the purple and red arrows, respectively. CMV, cytomegalovirus promoter/enhancer; IRES, internal ribosome entry site; pA, polyA signal. Lower panel: Factors essential for NHEJ of DSBs on chromosomal DNA.

Therefore, the efficiency of NHEJ can be assessed by monitoring EGFP production. In addition, the DSBs produced by *I-Sce*I and the subsequent NHEJ of the two broken DNA strands can be monitored by quantitative PCR.

Nucleotide sequencing of the joined DNA revealed that ligation required no (or very little) sequence homology between the DNA ends, indicating that the DNA ends were joined via NHEJ. In addition, the contribution of other factors essential for the NHEJ of incompatible DNA ends (whose involvement was indicated by *in vitro* and *in vivo* plasmid-based assays) was also confirmed in the present chromosome-based *in vivo* assay. These factors are KU80 and DNA-PKcs (synapsis), Artemis and PALF (DNA end resection), POLλ and POLμ (gap filling), and LIG4 (ligation) [21, 22].

2.3. HATs involved in NHEJ

To date, several distinct families of HAT proteins have been identified, including CBP, P300, PCAF, GCN5 and MYST (which includes TIP60) [23]. We investigated the effects of ablating *CBP, P300, PCAF* and *TIP60* on NHEJ using the cell-based NHEJ activity assay. A decrease in the number of EGFP-positive cells was observed upon transfection with siRNA specific for *CBP, P300* or *TIP60,* but not in cells transfected with siRNA specific for *PCAF* [6]. These results indicate that CBP, P300 and TIP60 are involved in NHEJ in human cells.

2.4. Natural compounds with HAT activity suppress NHEJ activity

Several compounds derived from natural ingredients show HAT-inhibitory activity (**Table 1**). Curcumin, a major curcumanoid found in the spice turmeric, is a specific inhibitor of the homologous HATs, CBP and P300 [24]. Anacardic acid, derived from the shell of *Anacardium occidentale* ('cashew nut'), inhibits P300, PCAF and TIP60 [7, 25, 26], and Garcinol, found in the rind of *Garcinia indica* (mangosteen), inhibits P300 and PCAF [27].

Compound	MW[1]	Target HATs	Other target proteins/pathways	DER[2]
Curcumin	368.38	CBP, P300	NF-κB pathway, PI3K/mTOR/ETS2 pathway, AP-1 STAT, LOX-1	1.23
Anacardic acid	342.47	P300, PCAF, TIP60	NF-κB pathway, LOX-1, Xanthine oxidase,	1.51
Garcinol	602.80	P300, PCAF	NF-κB pathway, Src, MAPK/ERK, PI3K/Akt pathways, topoisomerase I/II	1.61

[1]Molecular weight; [2]Dose enhancement ratio (as assessed in cell viability assays [8]).

Table 1. HAT inhibitors that suppress NHEJ in human cells.

The cell-based NHEJ activity assay was used to investigate the effects of curcumin, anacardic acid, and garcinol on NHEJ; the results showed that treatment with each compound decreased the proportion of EGFP-positive cells (**Figure 3**) [6, 8]. This indicates that these HAT inhibitors suppress NHEJ activity *in vivo*.

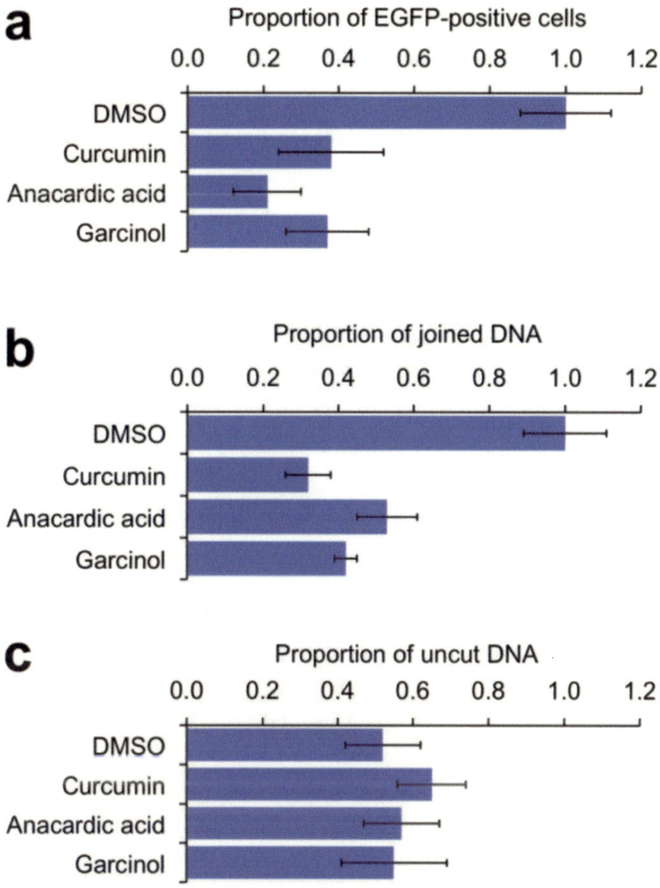

Figure 3. Suppression of NHEJ by HAT inhibitors [6, 8]. H1299-dA3-1 #1 cells pretreated with curcumin (20 μM), anacardic acid (50 μM), or garcinol (12 μM) were transfected with the *I-Sce*I expression plasmid. (a) The proportion of EGFP-positive cells assessed by fluorescence-activated cell sorting analysis. (b,c) Proportion of joined (b) and uncut (c) DNA in cells assessed by quantitative PCR. The proportion of EGFP-positive cells and the proportion of joined DNA after garcinol treatment versus those in cells treated with DMSO (expressed as a ratio). The proportion of uncut DNA remaining after drug treatment expressed as a ratio of the amount of uncut DNA present after *I-Sce*I transduction versus the amount present before transduction. Results 48 h after transfection of the *I-Sce*I expression plasmid are shown. The results represent the mean ± s.d from three independent experiments. siCTR, non-targeting siRNA.

2.5. HAT inhibitors radiosensitize cancer cells

Because DNA DSBs induced by IR are preferentially repaired by NHEJ [3, 4], and HAT inhibitors suppress the activity of NHEJ, it is thought that HAT inhibitors may enhance the intrinsic sensitivity of cancer cells to IR. In line with this, the radiosensitizing effects of curcumin, anacardic acid and garcinol have been studied by ourselves and others both *in vitro* and *in vivo* (see **Table 2**); however, it is not certain that the observed radiosensitizing effects of these compounds is entirely due to their HAT-inhibitory activity, since they may also affect many other proteins or pathways considered to be important for the cancer cell survival (see **Table 1**). In our own study, garcinol showed the strongest radiosensitization effect of the compounds tested. A nontoxic concentration of garcinol (4 uM) inhibited NHEJ without significantly affecting the DNA damage checkpoint **(Table 1)** [8]. Further investigations into mechanisms underlying the radiosensitizing effects of HAT inhibitors are ongoing.

Compound	Cells/mice	Cell lines	Authors	Year	Journal
Curcumin	Cells & mice	SCC1	Khafif A, *et al.*[9]	2009	The Laryngoscope
	Cells	HCT116	Sandur SK, *et al.*[10]	2009	Int J Radiat Oncol Biol Phys
	Cells	PC-3	Li M, *et al.*[11]	2007	Cancer Res
	Cells	PC-3	Chendil D, *et al.*[12]	2004	Oncogene
Curcumin /anacardic acid	Cells	H1299	Ogiwara H, *et al.*[6]	2011	Oncogene
Anacardic acid	Cells	SQ20B, SCC35, HeLa	Sun Y, *et al.*[7]	2006	FEBS Lett
Garcinol	Cells	A549, HeLa	Oike, *et al.*[8]	2012	Int J Radiat Oncol Biol Phys

Table 2. Radiosensitization by HAT inhibitors.

2.6. Clinical studies using compounds with HAT-inhibitory activity

There are several clinical studies reporting the administration of compounds with HAT-inhibitory to humans (**Table 3**). Curcumin has been used, either alone or combined with radiation therapy and/or chemotherapeutic agents, to treat cancer patients, and garcinol has been used for weight-loss therapy. Although not all of the studies were designed to specifically evaluate the radiosensitizing effects of these compounds, the data will be of help to estimate their toxicity. The available data indicate that the side effects of these compounds are tolerable, at least when used alone.

Compound	Disease	Phase	Modality	Sponsor/Author
Curcumin	Rectal cancer*	II	Curcumin, RT[1], capecitabine *vs.* RT[1], capecitabine	M.D. Anderson Cancer Center
Curcumin	Pancreatic cancer*	II	Curcumin, gemcitabine *vs.* gemcitabine	Rambam Healthcare Campus
Curcumin	Pancreatic cancer*	II	Curcumin alone	M.D. Anderson Cancer Center
Curcumin	Colorectal cancer	I	Curcumin alone	Sharma RA, et al.[13]
Curcumin	Healthy volunteer	I	Curcumin alone	Lao CD, et al.[14]
Garcinol	Obesity**	II	Garcinol, HCA[2] acid *vs.* HCA[2]	Sabinsa Corporation
Garcinol	Obesity	II	Garcinol, HCA[2], forskolin, piperine	Majeed M, et al.[15]

Details on the clinical trials are available for inspection at:
*http://www.clinicaltrials.gov/ and **http://www.garcitrin.com/clinical/.
[1]Radiation therapy, [2]Hydroxycitric acid.

Table 3. Clinical studies using compounds with HAT-inhibitory activity.

3. Conclusions/perspectives

The growing incidence of cancer worldwide indicates that radiation therapy will become increasingly significant as a cancer treatment [28]. Enhancing the efficacy of IR against cancer cells is urgent needs local control of tumors. As discussed in this article, radiosensitization of cancer cells by compounds with HAT-inhibitory activity has been reported at the level of basic research. Clinical studies indicate that some of these compounds can be administered to human patients with low systemic toxicity. Taken together, the available data suggest that compounds with HAT-inhibitory activity are promising candidates for radiosensitizers that may be applicable in clinical settings. However, the detailed mechanisms by which these compounds radiosensitize cancer cells are still largely unknown. Moreover, it is unclear whether these compounds can achieve adequate levels of radiosensitization in humans at a dose that shows no (or at least low) toxicity. Further investigations will establish whether HAT inhibitors can be used clinically to radiosensitize cancer cells.

Author details

Takahiro Oike
Division of Genome Biology, National Cancer Center Research Institute, Tokyo, Japan
Department of Radiation Oncology, Gunma University Graduate School of Medicine, Gunma, Japan

Hideaki Ogiwara and Takashi Kohno*
Division of Genome Biology, National Cancer Center Research Institute, Tokyo, Japan

Takashi Nakano
Department of Radiation Oncology, Gunma University Graduate School of Medicine, Gunma, Japan

Acknowledgement

This work was supported by Grants-in-Aid from the Ministry of Education, Culture, Sports, Science and Technology of Japan for Scientific Research on Innovative Areas (22131006) and from the Japan Society for the Promotion of Science for Young Scientists (B) KAKENHI (23701110); and the National Cancer Center Research and Development Fund.

4. References

[1] Hall EJ, Giaccia AJ. (2006) Radiosensitizers and bioreductive drugs. In: McAllister L, Bierig L, Barret K, editors. Radiobiology for the radiologist. Philadelphia: Lippincott Williams & Wilkins. pp. 419-431.

[2] Hall EJ, Giaccia AJ. (2006) DNA strand breaks and chromosomal aberrations. In: McAllister L, Bierig L, Barret K, editors. Radiobiology for the radiologist. Philadelphia: Lippincott Williams & Wilkins. pp. 16-29.

[3] Burma S, Chen BPC, Chen DJ, *et al.* (2006) Role of non-homologous end joining (NHEJ) in maintaining genomic integrity. DNA Repair. 8:1042-1048.

[4] Lieber MR. (2008) The mechanism of human nonhomologous DNA end joining. J. Biol. Chem. 283:1-5.

[5] Rossetto D, Truman AW, Kron SJ, *et al.* (2010) Epigenetic modifications in double-strand break DNA damage signaling and repair. Clin. Cancer. Res. 15:4543-4552.

[6] Ogiwara H, Ui A, Otsuka A, *et al.* (2011) Histone acetylation by CBP and p300 at double-strand break sites facilitates SWI/SNF chromatin remodeling and the recruitment of non-homologous end joining factors. Oncogene. 5:2135-2146.

[7] Sun Y, Jiang X, Chen S, *et al.* (2006) Inhibition of histone acetyltransferase activity by anacardic acid sensitizes tumor cells to ionizing radiation. FEBS Lett. 580:4353–4356.

[8] Oike T, Ogiwara H, Torikai K, *et al.* (2012) Garcinol, a histone acetyltransferase inhibitor, radiosensitizes cancer cells by inhibiting non-homologous end joining. Int. J. Radiat. Oncol. Biol. Phys. in press.

* Corresponding Author

[9] Khafif A, Lev-Ari S, Vexler A, *et al*. (2009) Curcumin: a potential radio-enhancer in head and neck cancer. Laryngoscope. 119:2019-2026.

[10] Sandur SK, Deorukhkar A, Pandey MK, *et al*. (2009) Curcumin modulates the radiosensitivity of colorectal cancer cells by suppressing constitutive and inducible NF-κB activity. Int. J. Radiat. Oncol. Biol. Phys. 75:534-542.

[11] Mao Li, Zhuo Zhang, Donald L. Hill, *et al*. (2007) Curcumin, a dietary component, has anticancer, chemosensitization, and radiosensitization effects by down-regulating the MDM2 oncogene through the PI3K/mTOR/ETS2 pathway. Cancer Res. 67:1988-96.

[12] Chendil D, Ranga RS, Meigooni D, *et al*. (2004) Curcumin confers radiosensitizing effect in prostate cancer cell line PC-3. Oncogene. 23:1599-1607.

[13] Sharma RA, Euden SA, Platton AL, *et al*. (2004) Phase I clinical trial of oral curcumin biomarkers of systemic activity and compliance. Clin. Cancer Res. 10:6847-1854.

[14] Lao CD, Ruffin MT, Normolle D, *et al*. (2006) Dose escalation of a curcuminoid formulation. BMC Complment Altern. Med. 6:10.

[15] Majeed M, Badmaev V, Khan N, *et al*. (2009) A new class of phytonutrients for body weight management. NUTRAfoods 8:17-26.

[16] Bird AW, Yu DY, Pray-Grant MG, *et al*. (2002) Acetylation of histone H4 by Esa1 is required for DNA double-strand break repair. Nature. 419:411–415.

[17] Tamburini BA, Tyler JK. (2005) Localized histone acetylation and deacetylation triggered by the homologous recombination pathway of double-strand DNA repair. Mol. Cell. Biol. 25:4903–4913.

[18] Murr R, Loizou JI, Yang YG, *et al*. (2006) Histone acetylation by Trrap-Tip60 modulates loading of repair proteins and repair of DNA double-strand breaks. Nat. Cell. Biol. 8:91–99.

[19] Lan L, Ui A, Nakajima S, Hatakeyama K, Hoshi M, Watanabe R, Janicki S, Ogiwara H, Kohno T, Kanno S, Yasui A. The ACF1 complex is required for DNA double-strand break repair in human cells. Mol Cell 2010, 40: 976-987.

[20] Jasin M. (1996) Genetic manipulation of genomes with rare-cutting endonucleases. Trends Genet. 12:224–228.

[21] Ogiwara H, Kohno T. Essential factors for incompatible DNA end joining at chromosomal DNA double strand breaks in vivo. PLoS One, 2011, 6(12) e28756.

[22] Li S, Kanno S, Watanabe R, Ogiwara H, Kohno T, Watanabe G, Yasui A, Lieber MR. PALF acts as both a single-stranded DNA endonuclease and a single-stranded DNA 3' -exonuclease and can participate in DNA end joining in a biochemical system. J Biol Chem. 2011,286:36368-77.

[23] Carrozza MJ, Utley RT, Workman JL, *et al*. (2003) The diverse functions of histone acetyltransferase complexes. Trends Genet. 19:321–329.

[24] Balasubramanyam K, Varier RA, Altaf M, *et al*. (2004) Curcumin, a novel p300/CREB-binding protein-specific inhibitor of acetyltransferase, represses the acetylation of histone/nonhistone proteins and histone acetyltransferase-dependent chromatin transcription. J. Biol. Chem. 279:51163-51171.

[25] Hemshekhar M, Sebastin SM, Kemparaju K, *et al.* (2011) Emerging roles of anacardic acid and its derivatives: a pharmacological overview. Basic Clin. Pharmacol. Toxicol. doi: 10.1111/j.1742-7843.2011.00833.x. [Epub ahead of print]

[26] Balasubramanyam K, Swaminathan V, Ranganathan A, *et al.* (2003) Small molecule modulators of histone acetyltransferase p300. J. Biol. Chem. 278:19134–19140.

[27] Balasubramanyam K, Altaf M, Varier RA, *et al.* (2004) Polyisoprenylated benzphenone, garcinol, anatural histone acetyltransferase inhibitor, represses chromatin transcription and alters global gene expression. J. Biol. Chem. 279:33716-33726.

[28] Boyle P. and Leon B. (2008) World Cancer Report 2008. Lyon: International agency for research on Cancer, World Health Organization. 15 p.

Chapter 2

A Framework for Modeling the Cellular Defending Mechanisms Against Genome Stress Under Radiotherapy

Jin-Peng Qi, Yong-Sheng Ding and Xian-Hui Zeng

Additional information is available at the end of the chapter

http://dx.doi.org/10.5772/56623

1. Introduction

Like immunotherapy, chemotherapy, and surgery, radiotherapy is one of the major tools in fighting against cancer. As acute IR is applied, cell can trigger its self-defensive mechanisms in response to genome stresses [1]. As one of the pivotal anticancer genes within the cell, P53 can control the transcription and translation of series genes, and trigger cell cycle arrest and apoptosis through interaction with downstream genes and their complicated signal pathways [2]. Under radiotherapy, the outcomes of cellular response depend on the presence of functional P53 proteins to induce tumor regression through apoptotic pathways [3]. Conversely, the P53 tumor suppressor is the most commonly known specific target of mutation in tumorigenesis [4]. Abnormalities in the P53 have been identified in over 60% of human cancers and the status of P53 within tumor cells has been proposed to be one of the determinant response to anticancer therapies [3,4]. Controlled radiotherapy studies show the existence of a strong biologic basis for considering P53 status as a radiation predictor [3,5]. Therefore, the status of P53 in tumor cell can be considered as a predictor for long-term biochemical control during and after radiotherapy [6-8].

Recently, several models have been proposed to explain the damped oscillations of P53 in cell populations [9-12]. However, the dynamic mechanism of the single-cell responses is not completely clear yet, and the complicated regulations among genes and their signal pathways need to be further addressed, particularly under the condition of acute IR.

Many studies have indicated that introducing novel mathematical and computational approaches can stimulate in-depth investigation into various complicated biological systems (see, e.g., [13-23]). These methods have provided useful tools for both basic research and drug development [24-33], helping understanding many marvelous action mechanisms in various biomacromolecular systems (see, e.g., [21,34-39]).

Based on the existing models [9-12] and inspired by the aforementioned mathematical and computational approaches in studying biological systems, here a new model is proposed for studying the P53 stress response networks under radiotherapy at the cellular level, along with the kinetics of DNA double-strand breaks (DSBs) generation and repair, ATM and ARF activation, as well as the regulating oscillations of P53-MDM2 feedback loop (MDM2 is an important negative regulator of the p53 tumor suppressor). Furthermore, the kinetics of the oncogenes degradation, as well as the eliminations of the mutation of P53 (mP53) and the toxins were presented. Also, the plausible outcomes of cellular response were analyzed under different IR dose domains.

It is instructive to mention that using differential equations and graphic approaches to study various dynamical and kinetic processes of biological systems can provide useful insights, as indicated by many previous studies on a series of important biological topics, such as enzyme-catalyzed reactions [18,40], low-frequency internal motions of biomacromolecules [41-46], protein folding kinetics [47,48], analysis of codon usage [49,50], base distribution in the anti-sense strands [51], hepatitis B viral infections [52], HBV virus gene missense mutation [53], GPCR type prediction [54], protein subcellular location prediction [55], and visual analysis of SARS-CoV [56,57].

In the present study, we are to use differential equations and directed graphic approaches to investigate the dynamic and kinetic processes of the cellular responding radiotherapy.

2. Method

2.1. Model review

Under the genome stresses, many efforts have been made to enhance P53-mediated transcription through some models [58,59] [9-12]. However, the interactions in a real system would make these models [60] extremely complicated. Therefore, a new feasible model is needed in order to incorporate more biochemical information. To realize this, let us take the following criteria or assumptions for the new model: **(1)** only the vital components and interactions are taken into account; **(2)** all the localization issues are ignored; **(3)** the simple linear relations are used to describe the interactions among the components concerned; and **(4)** there are enough substances to keep the system ''workable'' [58].

The new integrated model thus established for the P53 stress response networks under radiotherapy is illustrated in **Fig.1**. Compared with the previous models [9-12], the current model contains more vital components, such as oncogenes, ARF and mP53, as well as their related regulating pathways. In the DSBs generation and repair module, the acute IR induces DSBs stochastically and forms DSB-protein complexes (DSBCs) at each of the damage sites after interacting with the DNA repair proteins [2,3]. As a sensor of genome stress, ATM is activated by the DSBCs signal transferred from DSBs. Meanwhile, the over-expression of oncogenes prompted by acute IR can trigger the activation of ARF, further prompting the ATM activation [2] [7]. The cooperating effects of active ATM (ATM*) and active ARF (ARF*) switch on or off the P53-MDM2 feedback loop [2] [7,9], further regulating the downstream genes to control the cell cycle arrest and the cell apoptosis in response to genome stresses [8]. Here, we use the superscript * to represent the activate state as done in [61].

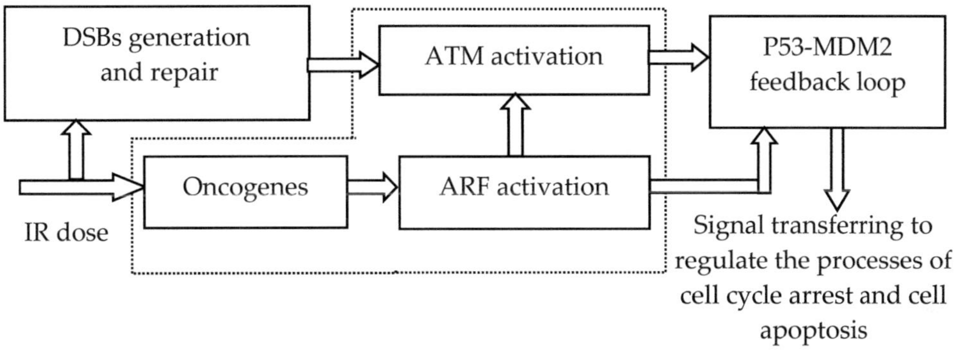

Figure 1. Illustration showing the integrated model of P53 stress response networks under radiotherapy. It is composed of three modules, including DNA damage generation and repair, ATM and ARF activation, as well as P53-MDM2 feedback loop. As acute IR is applied, ARF is activated by the over-expression of oncogenes, and ATM is activated with the cooperation of DSBCs and ARF*. ATM* and ARF* corporately trigger the responding mechanism of P53-MDM2 feedback loop.

2.2. DSBs generation and repair

Under the continuous effect of acute IR dose, DSBs occur and trigger two major repair mechanisms in eukaryotic cells: homologous recombination (HR) and nonhomologous end joining (NHEJ) [62,63]. About 60-80% of DSBs are rejoined quickly, whereas the remaining 20-40% of DSBs are rejoined more slowly [64,65]. As shown in **Fig.2**, the module of DSBs generation and repair process contains both the fast and slow kinetics, with each being composed of a reversible binding of repair proteins and DSB lesions into DSBCs, and an irreversible process from the DSBCs to the fixed DSBs [62,65]. DSBCs are synthesized by binding the resulting DSBs with repair proteins (RP), which is the main signal source to transfer the DNA damage to P53-MDM2 feedback loop by ATM activation [2].

Due to the misrepair part of DSBs (F_w) having the profound consequences on the subsequent cellular viability and the cellular response in fighting against genome stresses [1,3], we obviously distinguish between correct repair part of DSBs (F_r) and F_w [9,10,12]. Moreover, we further deal the total F_w in both repair processes as a part of toxins within the cell [2,4,11], which can be eliminated by the regulatory functions of P53 during and after radiotherapy, and treated as an indicator of outcomes in cellular response to genome stresses [2].

Some experimental data suggest that the quantity of the resulting DSBs within different IR dose domains obey a Poisson distribution [11]. In accordance with the experiments, we

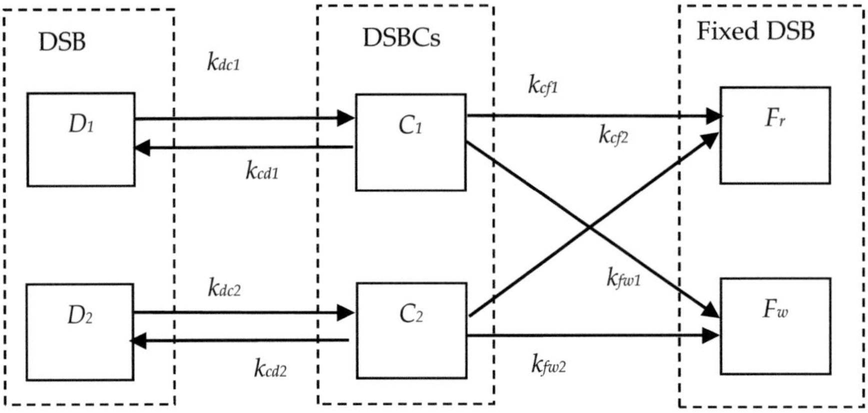

Figure 2. Illustration showing the module of DNA repair process. It includes both a fast repair pathway and a slow one. DSB can be in one of four states: intact DSB (DSB), DBSC, F_r and F_w. Subscripts '1' and '2' refer to the fast kinetics and slow one.

assume that the stochastic number of the resulting DSBs per time scale is proportional to the number generated by a Poisson random function during the period of acute radiation [11]. The DSBs generation process is formulated as follows:

$$\frac{d[\mathrm{DT}]}{dt} = k_t \times \mathrm{Poissrnd}(a_{ir} \times \mathrm{IR}) \tag{1}$$

where [DT] is the concentration of total resulting DSBs induced by IR in both fast and slow repair processes. k_t is the parameter to set the number of DSBs per time scale, and a_{ir} is the parameter to set the number of DSBs per IR dose.

Moreover, we assume that the limited repair proteins are available around DSBs sites, and 70% of the initial DSBs are fixed by the fast repair process. Each DSB can be in one of the four states: intact DSB, DSBC, F_r and F_w [9,10,12]. Thus, we have the following differential equations:

$$\frac{d[\mathrm{D}_1]}{dt} = a_1[\mathrm{D}_t] + k_{cd1}[\mathrm{C}_1] - [\mathrm{RP}](k_{\mathrm{dc1}}[\mathrm{D}_1] + k_{\mathrm{cross}}([\mathrm{D}_1] + [\mathrm{D}_2])) \tag{2}$$

$$\frac{d[\mathrm{D}_2]}{dt} = a_2[\mathrm{D}_t] + k_{\mathrm{cd2}}[\mathrm{C}_2] - [\mathrm{RP}](k_{\mathrm{dc2}}[\mathrm{D}_2] + k_{\mathrm{cross}}([\mathrm{D}_1] + [\mathrm{D}_2]) \tag{3}$$

$$\frac{d[\mathrm{C}_1]}{dt} = k_{\mathrm{dc1}}[\mathrm{D}_1] - k_{\mathrm{cd1}}[\mathrm{C}_1] - k_{\mathrm{cf1}}[\mathrm{C}_1] \tag{4}$$

$$\frac{d[\mathrm{C}_2]}{dt} = k_{\mathrm{dc2}}[\mathrm{D}_2] - k_{\mathrm{cd2}}[\mathrm{C}_2] - k_{\mathrm{cf2}}[\mathrm{C}_2] \tag{5}$$

$$\frac{d[\mathrm{RP}]}{dt} = S_{rp} + k_{\mathrm{cd1}}[\mathrm{C}_1] + k_{\mathrm{cd2}}[\mathrm{C}_2] - [\mathrm{RP}](k_{\mathrm{dc1}}[\mathrm{D}_1] + k_{\mathrm{dc2}}[\mathrm{D}_2] + k_{\mathrm{cross}}([\mathrm{D}_1] + [\mathrm{D}_2])) \tag{6}$$

$$\frac{d[\mathrm{F_w}]}{dt} = k_{fw1}[\mathrm{C_1}] + k_{fw2}[\mathrm{C_2}] \tag{7}$$

where [D], [C], and [F_w] represent the concentrations of DSBs, DSBCs, and F_w in the fast and the slow repair kinetics respectively, k_{dc}, k_{cd}, k_{cf}, and k_{fw} are the transition rates among the above three states; k_{dc}, and k_{cross} represent the first-order and second-order rate constants in both the fast and the slow repair kinetics respectively [65]. S_{rp} is the basal induction rate of repair mRNA, and subscripts '1' and '2' refer to the fast and the slow kinetics.

2.3. ATM and ARF activation

As a DNA damage detector, ATM exists as a dimer in unstressed cells. After IR is applied, intermolecular autophosphorylation occurs, causing the dimer to dissociate rapidly into the active monomers. The active ATM monomer (ATM*) can prompt the P53 expression further [64]. Meanwhile, ARF, another tumor suppressor, is activated by hyperproliferative signals emanating from oncogenes, such as Ras, c-myc etc., further prompting the ATM activation [2,7,10]. Based on the existing model of ATM switch [11], we present an ATM and ARF activation module under IR. Shown in **Fig.3** is the module scheme of ATM and ARF activation, which includes five components: ATM dimer, inactive ATM monomer, ATM*, ARF, and ARF*. Compared with the previous studies in [9-12], ARF, oncogenes, and the related signal pathways are involved in this module [2,7. Here, let us assume that DSBCs is the main signal transduction from DSBs to P53-MDM2 feedback loop through ATM activation, and the rate of ATM activation is a function of the amount of DSBCs, ARF* and the self-feedback of ATM*. Furthermore, the total concentration of ATM is a constant, including ATM dimer, ATM monomer and ATM, as treated in {Ma, 2005 #1194].

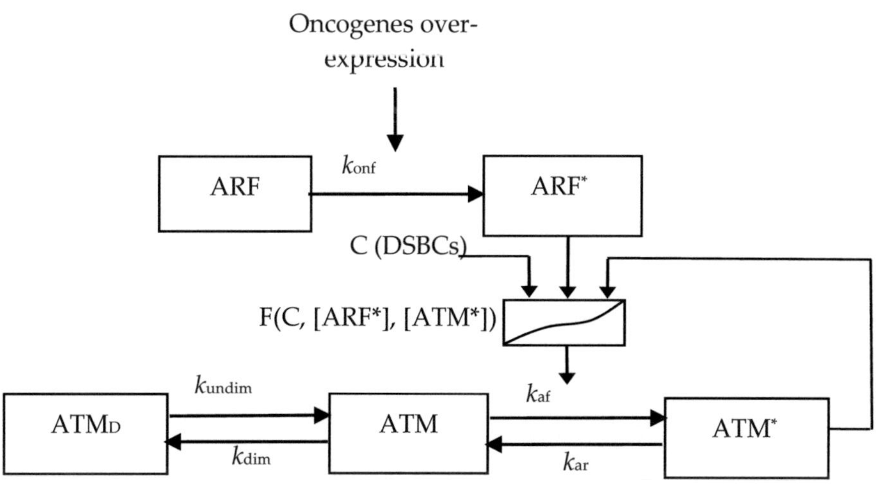

Figure 3. Illustration showing the module scheme of ATM and ARF activation under constant IR. ARF is activated by the over-expression of oncogenes induced by acute IR, and ATM is activated from ATM monomers under the cooperating effects of DSBCs, ARF*, and self-feedback of ATM*.

As a detector of DNA damage, ATM activation plays an important role in triggering the regulatory mechanisms of P53 stress response networks [2,65]. After the acute IR is applied, phosphorylation of inactive ATM monomers is promoted first by DSBCs and then rapidly by means of the positive feedback from ATM*, accounting for the intermolecular autophosphorylation [11]. Meanwhile, under the circumstance of continuous IR dose, ARF, a detector of over-expression of oncogenes is activated by hyperproliferative signals emanating from oncogenes, further prompting the ATM activation [2,7,10], as can be formulated as follows:

$$\frac{d[\mathrm{ATM_d}]}{dt} = \frac{1}{2}k_{\mathrm{dim}}[\mathrm{ATM_m}]^2 - k_{un\,\mathrm{dim}}[\mathrm{ATM_d}] \tag{8}$$

$$\frac{d[\mathrm{ATM_m}]}{dt} = 2k_{\mathrm{undim}}[\mathrm{ATM_d}] - k_{\mathrm{dim}}[\mathrm{ATM_m}]^2 - k_{\mathrm{af}}f[\mathrm{ATM_m}] + k_{\mathrm{ar}}[\mathrm{ATM^*}] \tag{9}$$

$$\frac{d[\mathrm{ATM^*}]}{dt} = k_{\mathrm{af}}f[\mathrm{ATM_m}] - k_{ar}[\mathrm{ATM^*}] \tag{10}$$

$$\frac{d[\mathrm{ARF}]}{dt} = S_{\mathrm{arf}} - k_{\mathrm{ad}}[\mathrm{ARF}] - k_{\mathrm{onf}}[\mathrm{Onco}][\mathrm{ARF}] \tag{11}$$

$$\frac{d[\mathrm{ARF^*}]}{dt} = k_{\mathrm{onf}}[\mathrm{Onco}][\mathrm{ARF}] - k_{\mathrm{pad}}[\mathrm{ARF^*}] \tag{12}$$

$$f(C,[\mathrm{ATM^*}]) = a_1C + a_2[\mathrm{ATM^*}] + a_3C[\mathrm{ATM^*}] + a_4[ARF^*] \tag{13}$$

where $[\mathrm{ATM_d}]$, [ATM] and [ATM*] represent the concentrations of ATM dimer, ATM monomer, and active ATM monomer respectively; [Onco], [ARF] and [ARF*] represent the concentrations of oncogenes, ARF, and active ARF respectively; k_{undim}, k_{dim}, k_{ar}, and k_{af} are the rates of ATM undimerization, ATM dimerization, ATM monomer inactivation, and ATM monomer activation, respectively. S_{arf}, k_{onf}, k_{ad} and k_{pad} are the rates of ARF basal induction, ARF activation triggered by Oncogenes, ARF degradation, and ARF* degradation, respetively. In addition, f is the function of ATM activation, the term a_1C implies the fact that DSBs somehow activate ATM molecules at a distance, a_2[ATM*] indicates the mechanism of autophosphorylation of ATM, a_3C[ATM*] represents the interaction between the DSBCs and ATM* [9-12,66], and a_4[ARF*] represents the regulating function of ARF* to ATM activation [1,3,7].

2.4. Regulation of P53-MDM2 feedback loop

As shown in **Fig.4**, P53 and its principal antagonist, MDM2 transactivated by P53, form a P53-MDM2 feedback loop, which is the core part in the integrated networks [9-12]. ATM* elevates the transcriptional activity of P53 by prompting phosphorylation of P53 and degradation of MDM2 protein [67]. Also, ARF* can indirectly prompt the transcriptional activity of P53 by inhibiting the expression of MDM2 and preventing P53 degradation [2,7,9]. With the cooperating regulations of ATM* and ARF*, this negative feedback loop can produce oscillations in response to the sufficiently strong IR dose [11].

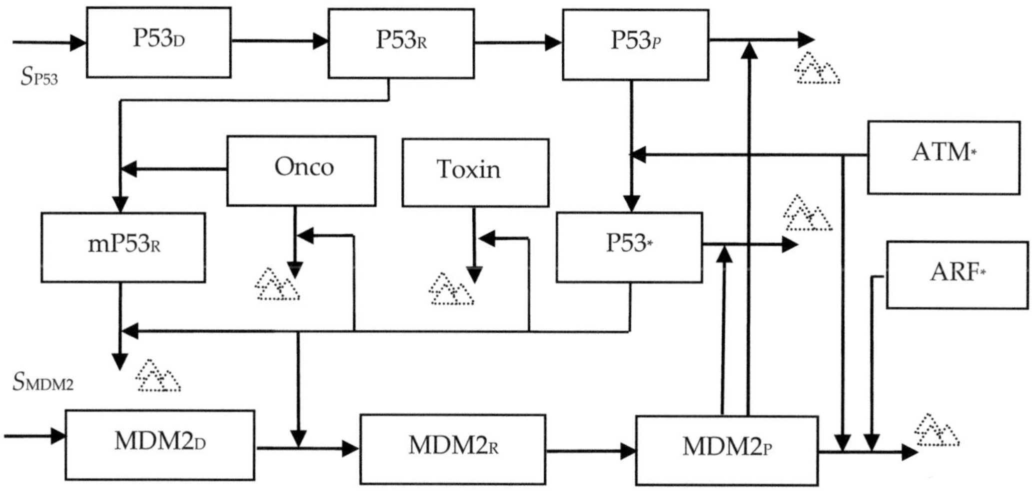

Figure 4. The directed graph of P53-MDM2 feedback loop under radiotherapy. P53 is translated from P53mRNA and phosphorylated by ATM* and ARF*. MDM2 protein promotes a fast degradation of P53 protein and a slow degradation of P53*. In addition, ATM*and ARF* stimulate the degradation of MDM2, and then indirectly increase the regulatory activation of P53* further. Especially, oncogenes, toxins and mP53 are decreased directly by the regulatory functions of P53*.

Especially, the mutation of P53 (mP53) triggered by oncogenes is added in this module, and mP53 is further dealt as another detector of outcomes in cellular response to acute IR. To account for a decreased binding affinity between inactive P53 and P53*, we assume that MDM2-induced degradation of inactive P53 is faster than that of P53*, and only P53* can induce target genes to depress the over-expression of oncogenes and further eliminate the toxins within the cell [3,4,9-12]. The main differential equations used in this module are as follows:

$$\frac{d[\mathrm{P53_R}]}{dt} = S_{\mathrm{P53}} - d_{\mathrm{rp}}[\mathrm{P53_R}] - k_{\mathrm{rp}}[\mathrm{P53_R}] \tag{14}$$

$$\frac{d[\mathrm{P53_P}]}{dt} = k_{\mathrm{rp}}[\mathrm{P53_R}] + k_{\mathrm{p^*p}}[\mathrm{P53^*}] - d_{\mathrm{pp}}[\mathrm{P53_P}]$$

$$-k_{\mathrm{app^*}}[\mathrm{ATM^*}]\frac{[\mathrm{P53_P}]}{[\mathrm{P53_P}]+k_{\mathrm{p}}} - k_{\mathrm{mp}}[\mathrm{MDM2_P}]\frac{[\mathrm{P53_P}]}{[\mathrm{P53_P}]+k_{\mathrm{d}}} \tag{15}$$

$$\frac{d[\mathrm{P53^*}]}{dt} = k_{\mathrm{app^*}}[\mathrm{ATM^*}]\frac{[\mathrm{P53_P}]}{[\mathrm{P53_P}]+k_{\mathrm{p}}} - k_{\mathrm{p^*p}}[\mathrm{P53^*}] - d_{\mathrm{pp^*}}[\mathrm{P53^*}]$$

$$-k_{\mathrm{p^*p}}[\mathrm{P53^*}]-d_{\mathrm{pp^*}}[\mathrm{P53^*}]-k_{\mathrm{mp^*}}[\mathrm{MDM2_P}]\frac{[\mathrm{P53^*}]}{[\mathrm{P53^*}]+k_{\mathrm{d^*}}} \tag{16}$$

$$\frac{d[\mathrm{MDM2_R}]}{dt}=S_{\mathrm{mdm2}}+k_{\mathrm{p^*m}}\frac{[\mathrm{P53^*}]^{\mathrm{n}}}{[\mathrm{P53^*}]^{\mathrm{n}}+k^{\mathrm{n}}}$$

$$-k_{\mathrm{mrp}}[\mathrm{MDM2_R}]-d_{\mathrm{mr}}[\mathrm{MDM2_R}] \tag{17}$$

$$\frac{d[\mathrm{MDM2_P}]}{dt}=k_{\mathrm{mrp}}[\mathrm{MDM2_R}]-d_{\mathrm{mp}}[\mathrm{MDM2_P}]$$

$$-(k_{\mathrm{mat}}\frac{[\mathrm{ATM^*}]}{[\mathrm{ATM^*}]+k_{\mathrm{at}}}+k_{\mathrm{mar}}\frac{[\mathrm{ARF^*}]}{[\mathrm{ARF^*}]+k_{\mathrm{ar}}})[\mathrm{MDM2_P}] \tag{18}$$

$$\frac{d[\mathrm{Onco}]}{dt}=k_{\mathrm{onIR}}[\mathrm{Onco}][\mathrm{IR}]-k_{\mathrm{onp}}[\mathrm{Onco}][\mathrm{P53^*}] \tag{19}$$

$$\frac{d[\mathrm{Toxins}]}{dt}=k_{\mathrm{tfw}}[\mathrm{F_w}]-k_{\mathrm{pt}}[\mathrm{P53^*}][\mathrm{Toxins}] \tag{20}$$

$$\frac{d[\mathrm{mP53}]}{dt}=k_{\mathrm{mp}}[\mathrm{P53_R}][\mathrm{Onco}]-k_{\mathrm{pmd}}[\mathrm{P53_{P^*}}][\mathrm{mP53}] \tag{21}$$

where $[P53_R]$, $[P53_P]$, $[P53^*]$, $[MDM2_R]$, and $[MDM2_P]$ represent the concentrations of P53 mRNA, P53 protein, active P53, MDM2 mRNA, and MDM2 protein, respectively; [Onco], [Toxins], and [mP53] represent the concentrations of oncogenes, F_w and mP53, respectively. S_{P53}, and S_{MDM2} represent the basal induction rates of P53 mRNA and MDM2 mRNA, respectively; k, and d represent the regulation and degradation rates among genes and proteins, respectively. The other parameters are presented in **Tables** 1-3.

Parameters	Description	Constant
k_t	Rate of DSBs generation per time scale	0.01
a_{ir}	Number of DSBs generation per IR dose	35
a_1	Percentage of DSs processed by fast repair	0.70
a_2	Percentage of DSs processed by slow repair	0.30
k_{dc1}	Rate of DSBs transition to DSBCs in fast repair process	2
k_{dc2}	Rate of DSBs transition to DSBCs in slow repair process	0.2
k_{dc1}	Rate of DSBCs transition to DSBs in fast repair process	0.5
k_{dc2}	Rate of DSBCs transition to DSBs in slow repair process	0.05
k_{fd1}	Rate of DSCs transition to F_d in fast repair process	0.001
k_{fd2}	Rate of DSCs transition to F_d in slow repair process	0.0001
k_{cross}	Rate of DSB binary mismatch in second order repair process	0.001

Table 1. The parameters used in the DSBs generation and repair processes

Parameters	Description	Constant
k_{dim}	ATM dimerization rate	8
k_{undim}	ATM undimerization rate	1
k_{af}	ATM phosphorylation rate	1
k_{ar}	ATM dephosphorylation rate	3
S_{arf}	Basal induction rate of ARF mRNA	0.001
k_{onf}	ARF activation rate triggered by Oncogenes	0.06
k_{ad}	ARF degradation rate	0.015
k_{pad}	ARF* degradation rate	0.01
a_1	Scale of the activation function of ATM phosphorylation	1
a_2	Scale of the activation function of ATM phosphorylation	0.08
a_3	Scale of the activation function of ATM phosphorylation	0.8

Table 2. The parameters used in the process of ATM and ARF activation

Parameters	Description	Constant
S_{P53}	Basal induction rate of P53 mRNA	0.001
d_{rp}	Degradation rate of P53 mRNA	0.02
k_{rp}	Translation rate of P53 mRNA	0.12
k_{p^*p}	Dephosphorylation rate of P53*	0.2
k_{app^*}	ATM*-dependent phosphorylation rate of P53	0.6
k_{mp}	MDM2-dependent degradation rate of P53	0.1
k_{mp^*}	MDM2-dependent degradation rate of P53*	0.02
d_{pp}	Basal degradation rate of P53	0.02
d_{pp^*}	Basal degradation rate of P53*	0.008
S_{MDM2}	Basal induction rate of MDM2 mRNA	0.002
k_{p^*m}	P53-dependent MDM2 transcription rate	0.03
k_{mrp}	Translation rate of MDM2 mRNA	0.02
d_{mr}	Degradation rate of MDM2 mRNA	0.01
d_{mp}	Basal degradation rate of MDM2	0.003
k_{mat}	ATM*-dependent degradation rate of MDM2	0.01
k_{mar}	ARF*-dependent degradation rate of MDM2	0.02
k_p	Michaelis constant of ATM*-dependent P53 phosphorylation	1.0
k	Michaelis constant of P53-dependent MDM2 transcription	1.0
k_d	Threshold concentration for MDM2-dependent P53 degradation	0.03
n	Hill coefficient of MDM2 transcription rate	4
k_{at}	Threshold concentration for ATM*-dependent MDM2 degradation	1.60
k_{ar}	Threshold concentration for ARF*-dependent MDM2 degradation	1.10
k_{d^*}	Threshold concentration for MDM2-dependent P53* degradation	0.32
k_{onIR}	Activation rate of oncogenes induced by IR	0.002
k_{onp}	Degredation rate of oncogenes induced by P53*	0.006

Parameters	Description	Constant
k_{tfw}	Toxins accumulation rate triggered by IR	0.6
k_{pt}	Toxins elimination rate triggered by P53*	0.1
k_{mpo}	Induction rate of mP53 induced by oncogenes over-expression	0.03
k_{mpd}	Elimination rate of mP53 triggered by P53*	0.015

Table 3. The parameters used in the process of P53-MDM2 loop and toxins degradation

3. Results and discussion

To ensure the accuracy of the simulation results, we consider that the valid parameter sets should obey the following rules [2,11,67]. **(1)** The model must contain oscillations because there has been experimental evidence that oscillations occur between P53 and MDM2 after cell stress. **(2)** The mechanism used to mathematically describe the degradation of P53 by MDM2 is accurate only for low concentrations of P53. **(3)** The concentration of P53* is much higher than that of inactive P53 after the system reaching an equilibrium.

Based on the above three rules and the existing parameter sets used in [11], we obtained the kinetics of P53 stress response networks and cellular response under acute IR dose through simulation platform in MATLAB7.0. The detailed parameters used for the current model are given in **Tables** 1-3.

3.1. Kinetics of DSBCs synthesizing

During the simulation process, the continuous 2, 5, and 7Gy IR are applied into a cell respectively. As shown in **Fig.5a**, owing to the condition that many DSBs occur and the limited RP are available around damage sites, the concentration of RP begins to decrease as IR dose overtakes 5Gy, and trends to zero versus radiation time. Meanwhile, the kinetics of DSBCs synthesizing is shown in **Fig.5b**. We can see that the rates of DSBCs synthesis keep increasing under 2, and 5Gy IR, whereas, it begins to decrease and trend to constant after about 120min under 7Gy IR dose.

3.2. Kinetics of ARF and ATM activation

The ARF activation is used to describe the mechanisms in cellular response to the over-expression of oncogenes induced by acute IR [2,7]. The kinetics of ARF activation is shown in **Fig.6a**. Owing to the over-expression of oncogenes without depressing functions of P53*, ARF is activated fast and ARF* keeps increasing followed by trending to dynamic equilibrium versus radiation time.

Meanwhile, the ATM activation module was established to describe the switch-like dynamics of the ATM activation in response to DSBCs increasing, and the regulation mechanisms during the process of the ATM transferring DNA damage signals to the P53-MDM2 feedback loop. Under the cooperative function of DSBCs, ARF*, and the positive self-feedback of ATM*, the ATM would reach the equilibrium state within minutes due to

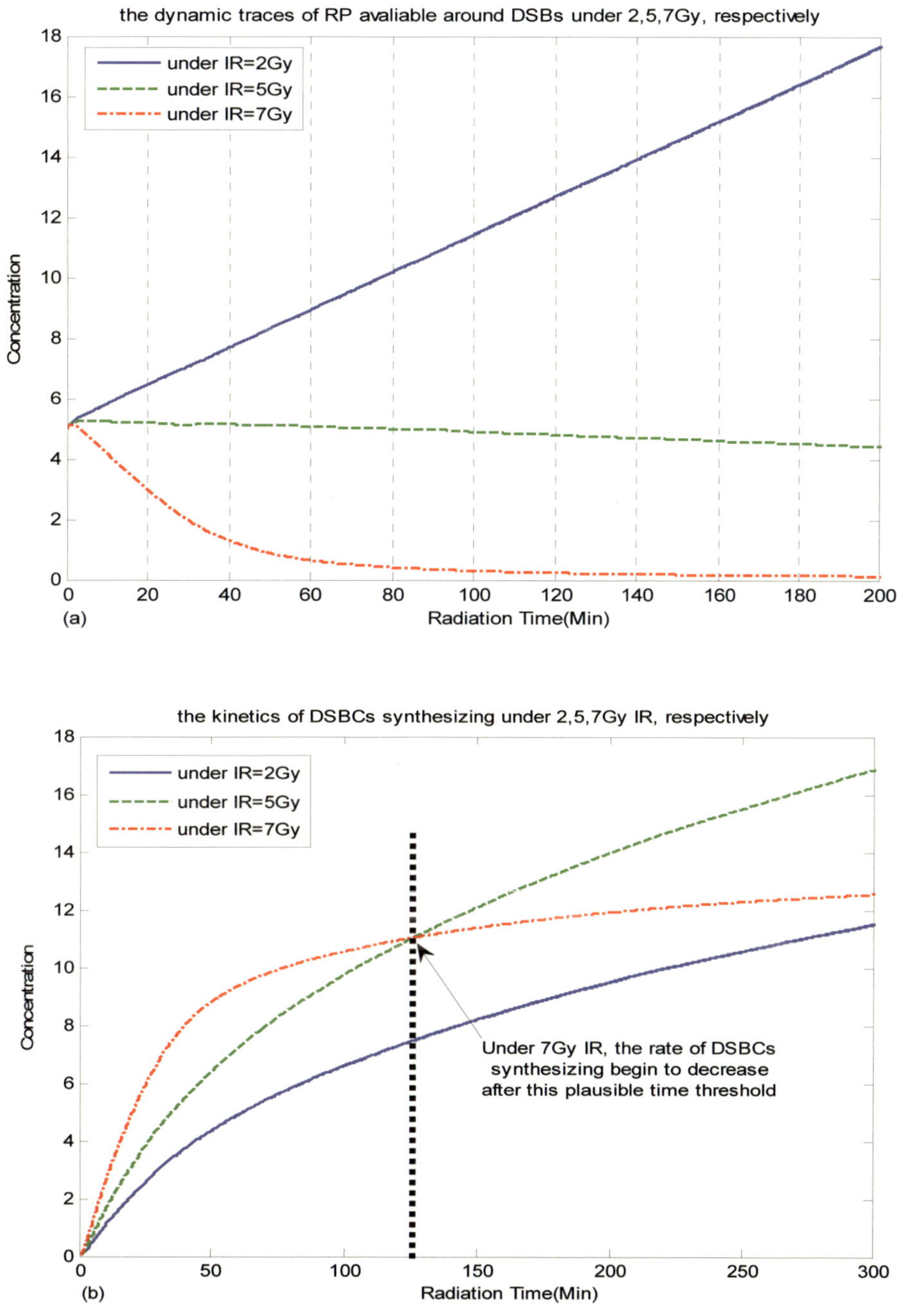

Figure 5. The kinetics of DSBs repairing and transferring under continuous effect of 2, 5, 7Gy IR.
(a) The dynamics of RP available around the resulting DSBs under different IR dose domains.
(b) The kinetics of DSBCs synthesized by DSBs and RP versus continuous radiation time under different IR dose domains.

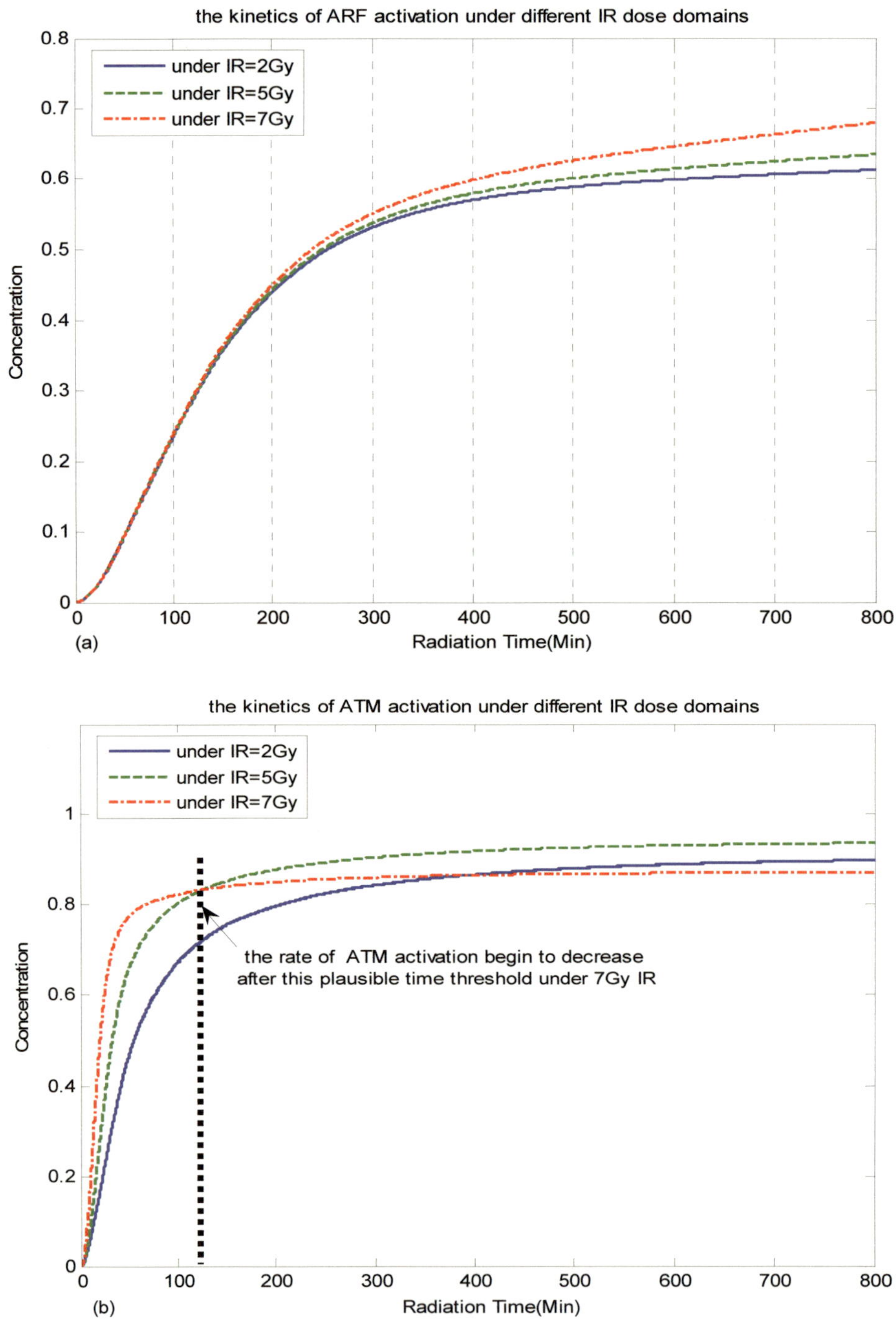

Figure 6. The kinetics of ARF and ATM activation under 2, 5, 7Gy IR. **(a)** The kinetics of ARF activation in response to over-expression of oncogenes induced by different IR dose. **(b)** The switch-like kinetics of ATM activation, ATM* reach saturation and trend to constant state in response to continuous radiation time of different IR dose domains.

the fast phosphorylation [2,11,67]. Kinetics of ATM activation is shown in **Fig.6b**. ATM is activated rapidly and switches to "on" state with respective rates, and then trends to the saturation state. The step-like traces suggest that the ATM module can produce an on-off switching signal, and transfer the damage signal to the P53-MDM2 feedback loop [3]. Furthermore, under the cooperation effects of ATM* and ARF*, DNA damage signals can be further transferred to the downstream genes and their signal pathways more efficiently [2,7].

3.3. Outcomes of cellular responding radiotherapy

The P53-MDM2 feedback loop is a vital part in controlling the downstream genes and regulation pathways to fight against the genome stresses [6,67,68]. In response to the input signal of ATM* and ARF*, the P53-MDM2 module generates one or more oscillations. The response traces of P53 and MDM2 protein under continuous application of 2, 5, and 7Gy IR from time 0 are shown in **Fig. 7a.** Upon the activation by ATM*, ARF* and decreased degradation by MDM2, the total amount of P53 proteins increases quickly. Due to the P53-dependent induction of MDM2 transcription, the increase of MDM2 proteins is sufficiently large to lower the P53 level, which in turn reduces the amount of the MDM2 proteins.

The oscillation pulses shown in **Fig.7a** have a period of 400 min, and the phase difference between P53 and MDM2 is about 100 min. Moreover, the first pulse is slightly higher than the second, quite consistent with the experimental observations [2,7,11] as well as the previous simulation results [9,10,12,69].

Also, by comparing these simulation results, we can see that the strength and swing of these oscillations begin to decrease as IR overtakes 7Gy, suggesting that the ability of cellular responding genome stresses begin to decrease as IR dose exceeds a certain threshold.

Furthermore, because in the current model the toxins, mP53 and oncogenes can be degraded directly by P53* in this module, we can plot the predictable outcomes of cellular response in fighting against genome stresses under different IR dose domains. As shown in **Fig.7b**, F_w remaining within the cell keeps decreasing with respective rate, and trends to zero versus continuous radiation time under 2 and 5Gy IR. Whereas, when IR exceeds 7Gy, F_w begins to increase slightly with some oscillations. Also, the kinetics of oncogenes degrading is plotted in **Fig.7c**. As we can see, owing to the negative regulations of P53*, the expression level of oncogenes keeps decreasing after the first climate under 2 and 5Gy IR dose, and then begins to increase slowly under 7Gy IR dose. Meanwhile, as shown in **Fig.7d**, quite similar to the results in **Fig.7b** and **Fig.7c**, mP53 keeps decrease after reaching the first maximum under 2 and 5Gy IR dose, and then begins to increase slowly under 7Gy IR dose. All these results obtained by the above simulations based on the new model indicate that that P53* indeed acts an important role in regulating downstream genes and their signal pathways, whereas its capabilities in cellular responding DNA damage under radiotherapy begin to decrease as the strength of IR exceeds a certain maximal threshold.

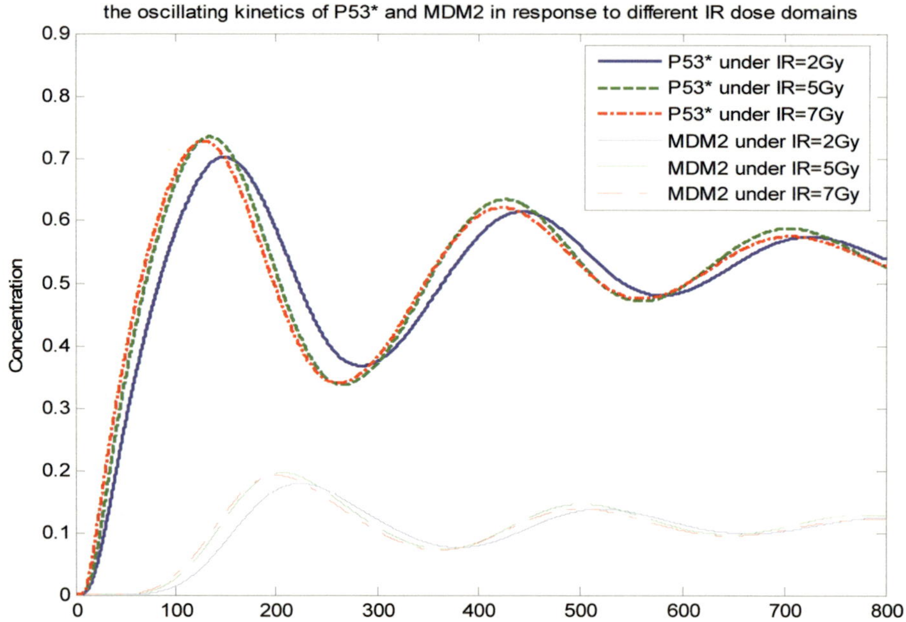
the oscillating kinetics of P53* and MDM2 in response to different IR dose domains
P53* under IR=2Gy
P53* under IR=5Gy
P53* under IR=7Gy
MDM2 under IR=2Gy
MDM2 under IR=5Gy
MDM2 under IR=7Gy
Concentration
Radiation Time(Min)
0.9
0.8
0.7
0.6
0.5
0.4
0.3
0.2
0.1
0
0
100
200
300
400
500
600
700
800

(a)

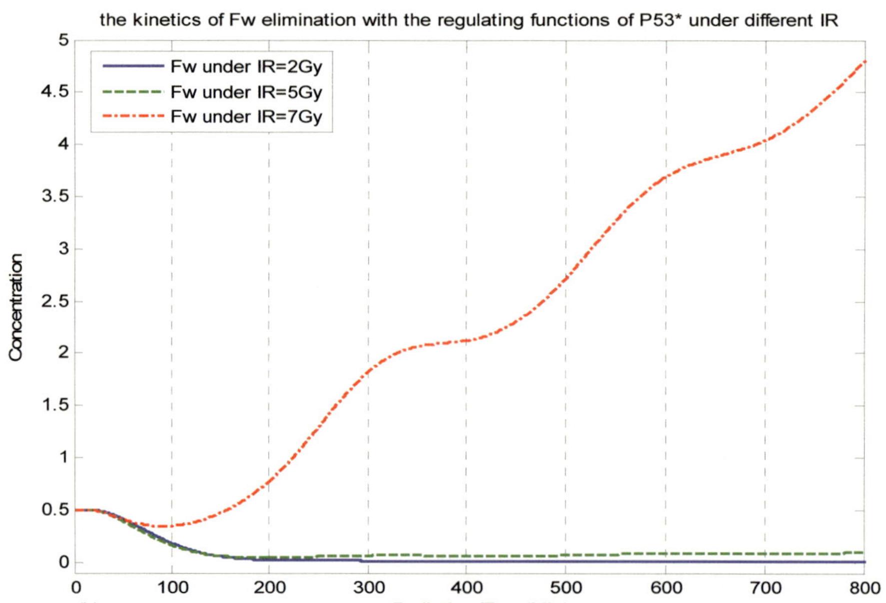
the kinetics of Fw elimination with the regulating functions of P53* under different IR
Fw under IR=2Gy
Fw under IR=5Gy
Fw under IR=7Gy
Concentration
Radiation Time(Min)
5
4.5
4
3.5
3
2.5
2
1.5
1
0.5
0
0
100
200
300
400
500
600
700
800

(b)

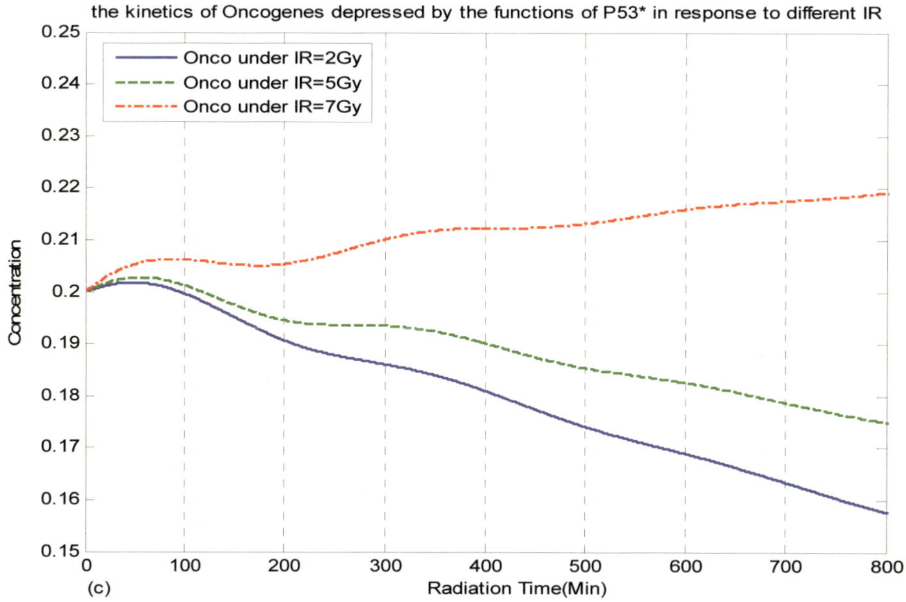

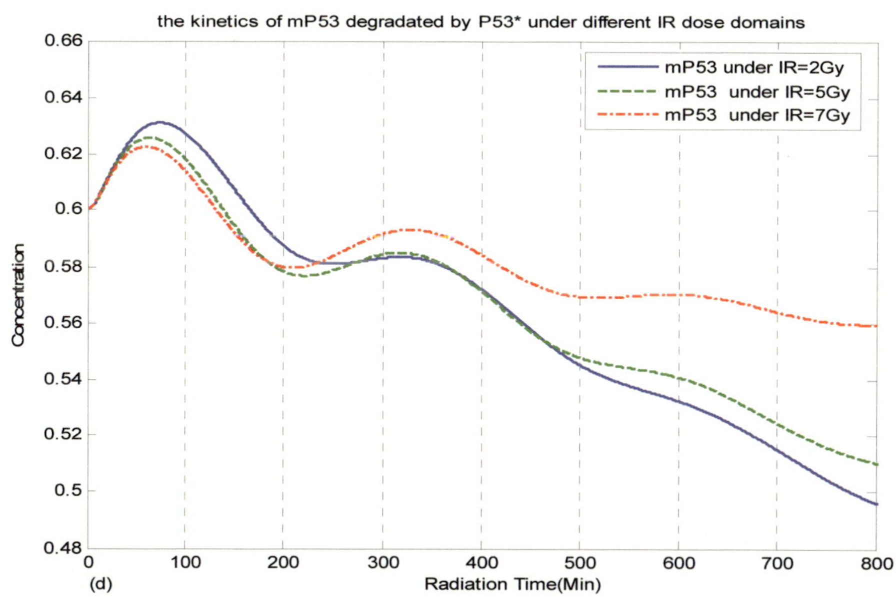

Figure 7. The outcomes of cellular responding 2, 5, 7Gy IR under radiotherapy. (a) The oscillating kinetics of P53* and MDM2 in response to the cooperative effect of ATM* and ARF* under different IR dose domains. (b) The kinetics of toxins elimination triggered by the functions of P53*. (c) The depressing dynamics of oncogenes over-expression with the regulations of P53*. (d) The kinetics of mP53 elimination triggered by the effect of P53*.

4. Conclusion

A new model was proposed to simulate the P53 stress response network under radiotherapy. It is demonstrated according to our model that ATM and ARF exhibits a strong sensitivity and switch-like behavior in response to the number of DSBs, fully consistent with the experimental observations. Interestingly, it is shown in this study that after the DNA damage signals transferring, P53-MDM2 feedback loop will produce oscillations, then triggering the cellular self-defense mechanisms to degrade the toxins remaining within the cell, such as F_w, oncogenes, and mP53. Particularly, under different IR dose domains, the new model can reasonably predict outcomes of cellular response in fighting against genome stresses, and hence providing a framework for analyzing the complicated regulations of P53 stress response networks, as well as the mechanisms of the cellular self-defense under radiotherapy.

Author details

Jin-Peng Qi*, Yong-Sheng Ding, Xian-Hui Zeng
College of Information Sciences and Technology Donghua University, Shanghai, China

Acknowledgement

The current work was supported in part by Specialized Research Fund for the Docto The current work was supported in part by Specialized Research Fund for the Doctoral Program of Higher Education from Ministry of Education of China (No. 20060255006), Project of the Shanghai Committee of Science and Technology (No. 08JC1400100), and the Open Fund from the Key Laboratory of MICCAI of Shanghai (06dz22103).ral Program of Higher Education from Ministry of Education of China (No. 20060255006), Project of the Shanghai Committee of Science and Technology (No. 08JC1400100), and the Open Fund from the Key Laboratory of MICCAI of Shanghai (06dz22103).

5. References

[1] Li, L., Story, M., Legerski, R.J. (2001) Cellular responses to ionizing radiation damage *Int J Radiat Oncol Biol Phys, 49,* 1157-1162.

[2] Kohn, K.W., Pommier, Y. (2005) Molecular interaction map of the p53 and Mdm2 logic elements, which control the Off-On switch of p53 in response to DNA damage *Biochem Biophys Res Commun, 331,* 816-827.

[3] Perez, C.A., Purdy, J.A. (1998) Treatment planning in radiation oncology and impact on outcome of therapy *Rays, 23,* 385-426.

[4] Tjebbes, G.W., Kreijveld, P.A., Tilanus, M.G., Hordijk, G.J., Slootweg, P.J. (2002) P53 tumor suppressor gene mutations in laryngeal cancer and in recurrent disease following radiation therapy *Oral Oncol, 38,* 296-300.

* Corresponding Author

[5] Duchting, W., Ulmer, W., Ginsberg, T. (1996) Cancer: a challenge for control theory and computer modelling *Eur J Cancer, 32A,* 1283-1292.

[6] Ritter, M.A., Gilchrist, K.W., Voytovich, M., Chappell, R.J., Verhoven, B.M. (2002) The role of p53 in radiation therapy outcomes for favorable-to-intermediate-risk prostate cancer *Int J Radiat Oncol Biol Phys, 53,* 574-580.

[7] Lindstrom, M.S., Wiman, K.G. (2003) Myc and E2F1 induce p53 through p14ARF-independent mechanisms in human fibroblasts *Oncogene, 22,* 4993-5005.

[8] Pauklin, S., Kristjuhan, A., Maimets, T., Jaks, V. (2005) ARF and ATM/ATR cooperate in p53-mediated apoptosis upon oncogenic stress *Biochem Biophys Res Commun, 334,* 386-394.

[9] Qi, J.P., Shao, S.H., Li, D.D., Zhou, G.P. (2007) A dynamic model for the p53 stress response networks under ion radiation *Amino Acids, 33,* 75–83.

[10] Qi, J.P., Shao, S.H., Xie, J., Zhu, Y. (2007) A mathematical model of P53 gene regulatory networks under radiotherapy *Biosystems, 90,* 698-706.

[11] Ma, L., Wagner, J., Rice, J.J., Hu, W., Levine, A.J., Stolovitzky, G.A. (2005) A plausible model for the digital response of p53 to DNA damage *Proc Natl Acad Sci U S A, 102,* 14266-14271.

[12] Qi, J.P., Shao, S.H., Shen, Y.Z. (2008) Cellular responding DNA damage: An improved modeling of P53 gene regulatory networks under ion radiation (IR) *Applied Mathematics and Computation, 205,* 73-83.

[13] Chou, K.C., Zhou, G.P. (1982) Role of the protein outside active site on the diffusion-controlled reaction of enzyme *Journal of American Chemical Society, 104,* 1409-1413.

[14] Chou, K.C. (1988) Review: Low-frequency collective motion in biomacromolecules and its biological functions *Biophysical Chemistry, 30,* 3-48.

[15] Chou, K.C. (1989) Graphical rules in steady and non-steady enzyme kinetics *J. Biol. Chem., 264,* 12074-12079.

[16] Chou, K.C. (1990) Review: Applications of graph theory to enzyme kinetics and protein folding kinetics. Steady and non-steady state systems *Biophysical Chemistry, 35,* 1-24.

[17] Ginsberg, T., Ulmer, W., Duchting, W. (1993) Computer simulation of fractionated radiotherapy: further results and their relevance to percutaneous irradiation and brachytherapy *Strahlenther Onkol, 169,* 304-310.

[18] Chou, K.C., Kezdy, F.J., Reusser, F. (1994) Review: Steady-state inhibition kinetics of processive nucleic acid polymerases and nucleases *Analytical Biochemistry, 221,* 217-230.

[19] Chou, K.C. (2004) Review: Structural bioinformatics and its impact to biomedical science *Current Medicinal Chemistry, 11,* 2105-2134.

[20] Chou, K.C. In *Structural bioinformatics and its impact to biomedical science and drug discovery, In: Frontiers in Medicinal Chemistry*; Atta-ur-Rahman and Reitz, A. B., Ed.; Bentham Science Publishers: The Netherlands, 2006; Vol. 3.

[21] Huang, R.B., Du, Q.S., Wang, C.H., Chou, K.C. (2008) An in-depth analysis of the biological functional studies based on the NMR M2 channel structure of influenza A virus *Biochem. Biophys. Res. Comm., 377,* 1243-1247.

[22] Chou, K.C., Shen, H.B. (2008) Cell-PLoc: A package of web-servers for predicting subcellular localization of proteins in various organisms *Nature Protocols, 3,* 153-162.

[23] Wang, J.F., Gong, K., Wei, D.Q., Li, Y.X., Chou, K.C. (2009) Molecular dynamics studies on the interactions of PTP1B with inhibitors: from the first phosphate binding site to the second one *Protein Engineering Design and Selection, 22*, 349-355.

[24] Althaus, I.W., Chou, J.J., Gonzales, A.J., Diebel, M.R., Chou, K.C., Kezdy, F.J., Romero, D.L., Aristoff, P.A., Tarpley, W.G., Reusser, F. (1993) Steady-state kinetic studies with the non-nucleoside HIV-1 reverse transcriptase inhibitor U-87201E *J. Biol. Chem., 268*, 6119-6124.

[25] Althaus, I.W., Gonzales, A.J., Chou, J.J., Diebel, M.R., Chou, K.C., Kezdy, F.J., Romero, D.L., Aristoff, P.A., Tarpley, W.G., Reusser, F. (1993) The quinoline U-78036 is a potent inhibitor of HIV-1 reverse transcriptase *J. Biol. Chem., 268*, 14875-14880.

[26] Althaus, I.W., Chou, J.J., Gonzales, A.J., Diebel, M.R., Chou, K.C., Kezdy, F.J., Romero, D.L., Aristoff, P.A., Tarpley, W.G., Reusser, F. (1993) Kinetic studies with the nonnucleoside HIV-1 reverse transcriptase inhibitor U-88204E *Biochemistry, 32*, 6548-6554.

[27] Chou, K.C., Wei, D.Q., Zhong, W.Z. (2003) Binding mechanism of coronavirus main proteinase with ligands and its implication to drug design against SARS. (Erratum: ibid., 2003, Vol.310, 675) *Biochem Biophys Res Comm, 308*, 148-151.

[28] Du, Q.S., Wang, S.Q., Jiang, Z.Q., Gao, W.N., Li, Y.D., Wei, D.Q., Chou, K.C. (2005) Application of bioinformatics in search for cleavable peptides of SARS-CoV Mpro and chemical modification of octapeptides *Medicinal Chemistry, 1*, 209-213.

[29] Zhang, R., Wei, D.Q., Du, Q.S., Chou, K.C. (2006) Molecular modeling studies of peptide drug candidates against SARS *Medicinal Chemistry, 2*, 309-314.

[30] Du, Q.S., Sun, H., Chou, K.C. (2007) Inhibitor design for SARS coronavirus main protease based on "distorted key theory" *Medicinal Chemistry, 3*, 1-6.

[31] Zheng, H., Wei, D.Q., Zhang, R., Wang, C., Wei, H., Chou, K.C. (2007) Screening for New Agonists against Alzheimer's Disease *Medicinal Chemistry, 3*, 488-493.

[32] Wei, H., Wang, C.H., Du, Q.S., Meng, J., Chou, K.C. (2009) Investigation into adamantane-based M2 inhibitors with FB-QSAR *Medicinal Chemistry, 5*, 305-317.

[33] Gong, K., Li, L., Wang, J.F., Cheng, F., Wei, D.Q., Chou, K.C. (2009) Binding mechanism of H5N1 influenza virus neuraminidase with ligands and its implication for drug design *Medicinal Chemistry, 5*, 242-249.

[34] Chou, K.C. (1987) The biological functions of low-frequency phonons: 6. A possible dynamic mechanism of allosteric transition in antibody molecules *Biopolymers, 26*, 285-295.

[35] Chou, K.C. (1989) Low-frequency resonance and cooperativity of hemoglobin *Trends in Biochemical Sciences, 14*, 212.

[36] Schnell, J.R., Chou, J.J. (2008) Structure and mechanism of the M2 proton channel of influenza A virus *Nature, 451*, 591-595.

[37] Gordon, G. (2008) Extrinsic electromagnetic fields, low frequency (phonon) vibrations, and control of cell function: a non-linear resonance system *Journal of Biomedical Science and Engineering (JBiSE), 1*, 152-156 (open accessible at http://www.srpublishing.org/journal/jbise/).

[38] Du, Q.S., Huang, R.B., Wang, C.H., Li, X.M., Chou, K.C. (2009) Energetic analysis of the two controversial drug binding sites of the M2 proton channel in influenza A virus *Journal of Theoretical Biology, 259,* 159-164.

[39] Pielak, R.M., Jason R. Schnell, J.R., Chou, J.J. (2009) Mechanism of drug inhibition and drug resistance of influenza A M2 channel *Proceedings of National Academy of Science, USA, 106,* 7379-7384.

[40] Chou, K.C., Jiang, S.P. (1974) Studies on the rate of diffusion-controlled reactions of enzymes *Scientia Sinica, 17,* 664-680.

[41] Chou, K.C. (1984) The biological functions of low-frequency phonons: 3. Helical structures and microenvironment *Biophysical Journal, 45,* 881-890.

[42] Chou, K.C., Kiang, Y.S. (1985) The biological functions of low-frequency phonons: 5. A phenomenological theory *Biophysical Chemistry, 22,* 219-235.

[43] Zhou, G.P. (1989) Biological functions of soliton and extra electron motion in DNA structure *Physica Scripta, 40,* 698-701.

[44] Chou, K.C., Maggiora, G.M., Mao, B. (1989) Quasi-continuum models of twist-like and accordion-like low-frequency motions in DNA *Biophysical Journal, 56,* 295-305.

[45] Chou, K.C., Zhang, C.T., Maggiora, G.M. (1994) Solitary wave dynamics as a mechanism for explaining the internal motion during microtubule growth *Biopolymers, 34,* 143-153.

[46] Sinkala, Z. (2006) Soliton/exciton transport in proteins *J Theor Biol, 241,* 919-927.

[47] Chou, K.C., Shen, H.B. (2009) FoldRate: A web-server for predicting protein folding rates from primary sequence *The Open Bioinformatics Journal, 3,* 31-50 (openly accessible at http://www.bentham.org/open/tobioij/).

[48] Shen, H.B., Song, J.N., Chou, K.C. (2009) Prediction of protein folding rates from primary sequence by fusing multiple sequential features *Journal of Biomedical Science and Engineering (JBiSE), 2,* 136-143 (openly accessible at http://www.srpublishing.org/journal/jbise/).

[49] Chou, K.C., Zhang, C.T. (1992) Diagrammatization of codon usage in 339 HIV proteins and its biological implication *AIDS Research and Human Retroviruses, 8,* 1967-1976.

[50] Zhang, C.T., Chou, K.C. (1994) Analysis of codon usage in 1562 E. Coli protein coding sequences *Journal of Molecular Biology, 238,* 1-8.

[51] Zhang, C.T., Chou, K.C. (1996) An analysis of base frequencies in the anti-sense strands corresponding to the 180 human protein coding sequences *Amino Acids, 10,* 253-262.

[52] Xiao, X., Shao, S.H., Chou, K.C. (2006) A probability cellular automaton model for hepatitis B viral infections *Biochem. Biophys. Res. Comm., 342,* 605-610.

[53] Xiao, X., Shao, S., Ding, Y., Huang, Z., Chen, X., Chou, K.C. (2005) An Application of Gene Comparative Image for Predicting the Effect on Replication Ratio by HBV Virus Gene Missense Mutation *Journal of Theoretical Biology, 235,* 555-565.

[54] Xiao, X., Wang, P., Chou, K.C. (2009) GPCR-CA: A cellular automaton image approach for predicting G-protein-coupled receptor functional classes *Journal of Computational Chemistry, 30,* 1414-1423.

[55] Xiao, X., Shao, S.H., Ding, Y.S., Huang, Z.D., Chou, K.C. (2006) Using cellular automata images and pseudo amino acid composition to predict protein subcellular location *Amino Acids, 30,* 49-54.

[56] Wang, M., Yao, J.S., Huang, Z.D., Xu, Z.J., Liu, G.P., Zhao, H.Y., Wang, X.Y., Yang, J., Zhu, Y.S., Chou, K.C. (2005) A new nucleotide-composition based fingerprint of SARS-CoV with visualization analysis *Medicinal Chemistry, 1,* 39-47.

[57] Gao, L., Ding, Y.S., Dai, H., Shao, S.H., Huang, Z.D., Chou, K.C. (2006) A novel fingerprint map for detecting SARS-CoV *Journal of Pharmaceutical and Biomedical Analysis, 41,* 246-250.

[58] Tyson, J.J. (1999) Models of cell cycle control in eukaryotes *J Biotechnol, 71,* 239-244.

[59] Tyson, J.J., Novak, B. (2001) Regulation of the eukaryotic cell cycle: molecular antagonism, hysteresis, and irreversible transitions *J Theor Biol, 210,* 249-263.

[60] Magne, N., Toillon, R.A., Bottero, V., Didelot, C., Houtte, P.V., Gerard, J.P., Peyron, J.F. (2006) NF-kappaB modulation and ionizing radiation: mechanisms and future directions for cancer treatment *Cancer Lett, 231,* 158-168.

[61] Chou, K.C., Watenpaugh, K.D., Heinrikson, R.L. (1999) A Model of the complex between cyclin-dependent kinase 5 (Cdk5) and the activation domain of neuronal Cdk5 activator *Biochemical & Biophysical Research Communications, 259,* 420-428.

[62] Rapp, A., Greulich, K.O. (2004) After double-strand break induction by UV-A, homologous recombination and nonhomologous end joining cooperate at the same DSB if both systems are available *J Cell Sci, 117,* 4935-4945.

[63] Rothkamm, K., Kruger, I., Thompson, L.H., Lobrich, M. (2003) Pathways of DNA double-strand break repair during the mammalian cell cycle *Mol Cell Biol, 23,* 5706-5715.

[64] Budman, J., Chu, G. (2005) Processing of DNA for nonhomologous end-joining by cell-free extract *EMBO J, 24,* 849-860.

[65] Daboussi, F., Dumay, A., Delacote, F., Lopez, B.S. (2002) DNA double-strand break repair signalling: the case of RAD51 post-translational regulation *Cell Signal, 14,* 969-975.

[66] Oren, M. (2003) Decision making by p53: life, death and cancer *Cell Death Differ, 10,* 431-442.

[67] Lev Bar-Or, R., Maya, R., Segel, L.A., Alon, U., Levine, A.J., Oren, M. (2000) Generation of oscillations by the p53-Mdm2 feedback loop: a theoretical and experimental study *Proc Natl Acad Sci U S A, 97,* 11250-11255.

[68] Weller, M. (1998) Predicting response to cancer chemotherapy: the role of p53 *Cell Tissue Res, 292,* 435-445.

[69] Lahav, G., Rosenfeld, N., Sigal, A., Geva-Zatorsky, N., Levine, A.J., Elowitz, M.B., Alon, U. (2004) Dynamics of the p53-Mdm2 feedback loop in individual cells *Nat Genet, 36,* 147-150.

Section 2

Dosimetry and Medical Physics

Applications to Radiotherapy Using Three Different Dosimetric Tools: MAGIC-*f* Gel, PENELOPE Simulation Code and Treatment Planning System

Thatiane Alves Pianoschi and Mirko Salomón Alva-Sánchez

Additional information is available at the end of the chapter

http://dx.doi.org/10.5772/56349

1. Introduction

Electron beams are widely used in radiotherapy with superior advantages in the irradiation of near-surface targets compared to photon beams, due to their characteristic therapeutic range and the plateau of the dose, finding between 80% and 90% of the maximum dose on central axis, and steep falloff of the dose with depth, characteristics that not exist in photon beams.

Thus, the electron beams are an important therapeutic modality for superficial treatments involving: skin and lip cancer, cancer of the chest wall and neck (after surgery and for recurrent cancers), upper respiratory and digestive tract lesions from 1 to 5 cm depth and reinforcement in the treatment of lymph nodes, scars from surgeries and residual tumors [1].

The main dosimetric parameter used for planning in radiotherapy with electron beams is obtained through the curves of percentage depth dose (PDD) [2]. From the PDD one can determine the maximum, practical and therapeutic range of the beam, the depth of maximum dose and depths that receive 90 % and 50% of the maximum dose.

Measurements of the dosimetric parameters with electron beam are more complex due to beam characteristics, especially the high dose gradient, which is present when the dose suffers a sharp drop after the build-up region. Standard dosimeters like ionization chamber, TLD and film do not have a high resolution, low energy dependence and the possibility of use with high dose gradient. Thus, the choice of the dosimeter for this type of beam is primordial.

The dosimetry gel have a high resolution, with atomic number equivalent to water and the possibility of providing measurement of high dose gradients in three-dimensions. Amongst the gel dosimeters, the MAGIC-*f* gel has been showed great concordance with the reference dosimeters.

After being exposed to ionizing radiation, the compounds of the MAGIC-*f* gel undergo a polymer reaction, that results in a chain of polymers that is completed after some days. The formation of the polymeric chain can be co-related with the absorbed dose, that can be seen on magnetic resonance images and through this imaging a three dimensional dose in target volume can be computed.

Another effective dosimetric tool for the study of this beam is the Monte Carlo simulation codes, that offers a convenient alternative compared to experimental methods, with advantage of providing detailed studies, and in different conditions that involve experimental procedures which are lengthy, complex and expensive [3]. The use of PENELOPE-Monte Carlo simulation code to simulate phenomena of attenuation of the dose radiation and dose deposition has been on an increase. The reliability of the results found by this code is directly related to the accuracy of transport models and the cross section libraries of the particles transported [4].

This chapter will be discuss the application of the two dosimetric tools, the MAGIC-*f* gel dosimeter and PENELOPE-Monte Carlo simulation code with high spatial resolution for determination of tridimensional dose distributions in target volumes for electron beams.

2. MAGIC-*f* gel dosimeter

Dosimeters based on polymeric gels are compounds that polymerize when subjected to radiation, this polymerization is related with the absorbed dose. Due to this property, these dosimeters have the ability to store information of the dose distributions in three-dimensios (3D). This is an advantage compared to other dosimeters providing only dose in a point or two-dimensional, as ionization chambers and films, respectively. This advantage is particularly important for the new technologies related with the radiation, where a significant incidence of high dose gradients is recorded.

The proposed sensitivity of gels to radiation was suggested by Stein and Day in 1950 when it was shown that the gels alter color depending on the absorbed dose [5]. In 1957 Andrews and colleagues studied the dose distribution and measurements of the pH of sensitive gels by spectroscopy [6]. The use of these gels as a dosimeter began with Gore and colleagues in 1984 when it was investigated the Fricke gels, initially studied by Fricke and Morse in 1927, based on the principle of oxidation, and recorded the relaxation properties in nuclear magnetic resonance (NMR) and showed that the concentration of ferric ions could be quantified by this technique [7].

Besides the research on Fricke gel, the studies with other gel dosimeters, polymer dosimeters as: BANANA (bis acrylamide and agarose nitrous oxide) [8], BANG (bis acrylamide gel nitrous oxide) [9] and PAG (acrylamide polymer gelatine) were started [10].

In polymeric dosimeters, monomeric compounds of the dosimeters are immersed in a gelatinous matrix, aqueous polymer suffer a reaction to the absorbed dose, resulting in a polymer gel matrix. This formation of radio-induced products changes the NMR relaxation properties, which can be related to the absorbed dose deposition, thus presenting a potential dosimeter for clinical dosimetry in 3D. However, the polymerization can be inhibited due to presence of oxygen, hence hypoxic conditions are required for its manufacture. To solve this problem Fong et al [11] created a new polymer gel, MAGIC (methacrylic and ascorbic acid in gelatin initiated by copper), formed by the combination of methacrylate-based materials, ascorbic acid and salt copper. The oxygen uptake is given by ascobato-copper complex, which allows the preparation of polymeric gels in normal atmospheric conditions in 2001[12-14]. Another problem presented by the polymer gels was the melting of the samples when stored at room temperature causing loss of information about dose distribution thereby restricting its use. In 2008, Fernandes and colleagues [15] solved this problem by adding formaldehyde to the original formulation of the MAGIC increasing its melting point to 69 ° C, and named the new gel MAGIC-f.

3. PENELOPE simulation code

The Monte Carlo method is a technique that uses the sampling of random numbers and statistical methods to find solutions to mathematical or physical problems [16]. In the Monte Carlo simulation (SMC) of radiation transport, the history of a particle is described as a probabilistic sequence of interactions when the particle changes its direction of movement, losing a part or all its energy, and occasionally generating a secondary particle [4].

Among the SMC codes used to simulate the interaction of radiation with matter, EGS [17], MCNP [18] and, more recently, PENELOPE [19] and GEANT [20] have been applied to radiology. The quality of the results provided by different simulation codes is directly linked with the accuracy of the transport model and implemented by libraries that contain the data associated with the cross section of particles transported [4]. The transport algorithm implemented by PENELOPE [3], led to its extensive use in radiotherapy [21-27].

Thus, the Monte Carlo simulation code PENELOPE, freely distributed by the Nuclear Energy Agency (NEA) is used to simulate the transport of electrons, positrons and photons in a complex geometry and an arbitrary material. The subroutines of FORTRAN code are organized into four basic files: PENELOPE.f containing the subroutines of transport of particles, PENGEOM.f containing subroutines geometry; PENVARED.f containing the subroutines that perform the methods of reducing variations and TIMER.f, which manages the simulation time. Besides these files, the code has a database with the characteristics of various materials of interest in radiological physics [28] cross section libraries and other quantities necessary for the transport of particles. One of the main advantages of using the code SMC is the use of recent cross-section libraries, EPDL97 [19].

The algorithm uses a simulation model PENELOPE combining numerical data and analytical cross section for the different types of interactions. It is applied from 1 keV energy to approximately 1 GeV where a detailed transport of photons is simulated by a conventional method. The simulation of electrons and positrons is made by means of a mixed algorithm because the latter undergo a large number of iterations before being effectively absorbed by the medium, resulting in small energy losses making it impractical to use a detailed method (or class I) for the transport of these particles.

Thus, for electrons and positrons, the PENELOPE code differs from other simulation codes by using a mixed algorithm (or class II), which implements two simulation models: a detailed, strong events, defined as the deflection angle (angle scattering) or loss of energy above a preset value, and condensed to weak interactions, with angular deflection (scattering angle) or loss of energy lower than the pre-set values. The condensed interactions are described by an approximation of multiple scattering, which consists in transforming a large number of weak interactions in a single artificial event. The multiple scattering theory algorithms implemented in the simulation is made condensed approximations and can lead to systematic errors assigned to the dependence of the simulation parameters that control the transport.

4. Treatment planning system

Actually every service of radiotherapy uses a treatment planning system (TPS) to plan a simulated irradiation for external or internal beam for a patient with some cancer, the manipulation of TPS is under the responsibility of the oncologist and the medical physicists, who try to minimize the dose in healthy structures and conform the dose in the tumor [29].

The calculated algorithms, which are based the TPS, use medical imaging from the patient obtained through technical images like: computed tomography, magnetic resonance imaging and positron emission tomography [30]. Today, the modern TPS provide tools for multimodality image matching, also known as image coregistration or fusion. Different dose prediction models are available, including pencil beam, cone beam and Monte Carlo simulation, with precision versus computation time being the relevant trade-off.

The treatment simulation is used to plan the geometric and radiological aspects of the therapy using radiation. Medical physicists plan the simulation treatment based on the prescribed dose stipulated by the oncologist and the constrains of the risk organs. Thus, the TPS is used to place beams which can deliver enough radiation to a tumor trying both the criteria: minimizing the dose to healthy tissue and risk organs and deliver the prescribed dose to the tumor. For this determination many decisions are to be considered including radiation beam (that are generally photons or electrons beams), angles of radiation incidence, irradiation field, whether attenuation wedges are to be used, and which multileaf collimator configuration will be used to shape the radiation from each beam [31]. Plans are often evaluated through dose-volume histograms, that can show the uniformity of the dose to the diseased tissue (tumor) and sparing of healthy structures. The obtained plan from the TPS can be evaluated comparing it with experimental measurements and also through the one simulation code.

5. MAGIC-*f*, PENELOPE and TPS use for dosimetry in some clinical cases for electron beams

5.1. Dosimetric response of the MAGIC-*f* gel for electron beams

Polymer gel dosimeters have been studied for use in dosimetry for photon beams for the characteristics of high spatial resolution and determination of dose in three-dimensional dose distributions. Some properties like response dependence on dose, energy and dose rate are not well established for electron beams.

The objective of this work is to evaluate the use of MAGIC-*f* gel dosimeter for electron beam in radiotherapy.

5.1.1. Materials and methods

Samples of MAGIC-*f* gel were manufactured following the protocols establish by Fernandes [15] and poured into three cylindrical glass tubes routinely used for blood sample collection (BD Vacutainer®) with 5ml volume, 12mm diameter for a specific measurement. Experimental irradiations were made at Hospital de Câncer de Barretos (HCB), using a Varian 2100c linear accelerator.

Variation of dose-response from Magic-*f* gel was evaluated verifying the possibility of the linear behavior of the gel for two energies, 9 and 15 MeV at a dose range of 1 to 10 Gy. To evaluate the response the dose rate were varied from 80cGy/min to 400cGy. The assessment of the response of the dosimeter in different depth was performed through the percentage depth dose (PDD) for the same energy with a irradiation field of 15 x 15 cm^2 at 100 cm from the water.

The readings of the gel samples were performed with the relaxometry technique in tomography mode of nuclear resonance magnetic (NMR), Philips 3.0 Tesla, from the section of Radiological from Hospital Clinic. The acquisition sequence of the NMR images were made with the multi spin-echo with 5 echoes, time echo of 20ms, repetition time and 0,250m spatial resolution. Figure 1 shows the NMR images and their maps of R2.

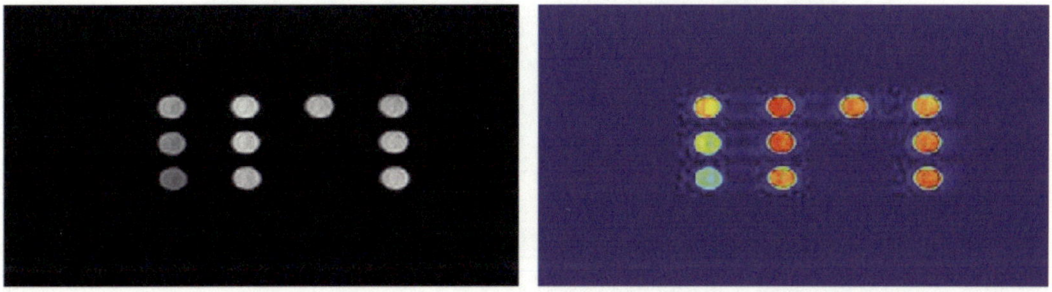

(a) (b)

Figure 1. Images of the axial section of the phantom: (a) NMR images; (b) R2 maps.

Figure 2 shows NMR images of relaxometry and R2 map normalized corresponding to the dose maps, when can be determine the PDD.

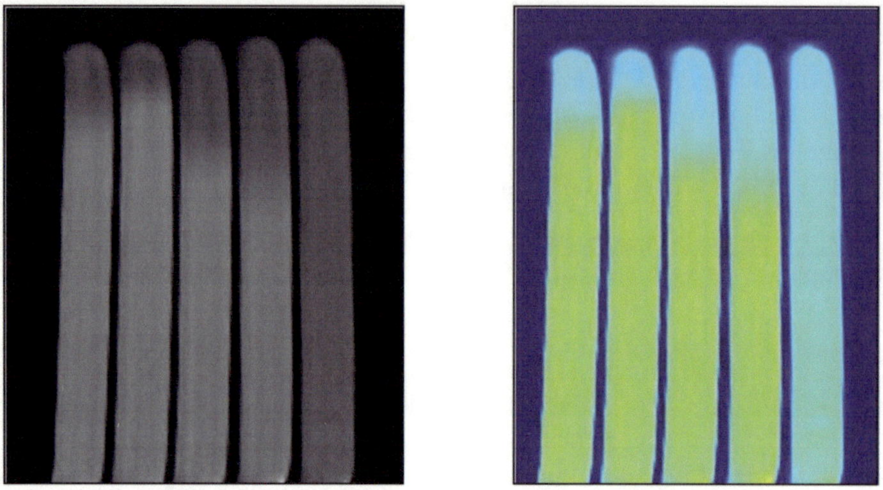

(a) (b)

Figure 2. Phantom to determine the PDD: (a) RNM image and (b) R2 maps.

5.1.2. Results and discussions

The irradiation with different dose rates have different degrees of polymerization, which can be visualized by the difference in tone of the phantoms irradiated, so that Figure 3 shows this difference of polymerization.

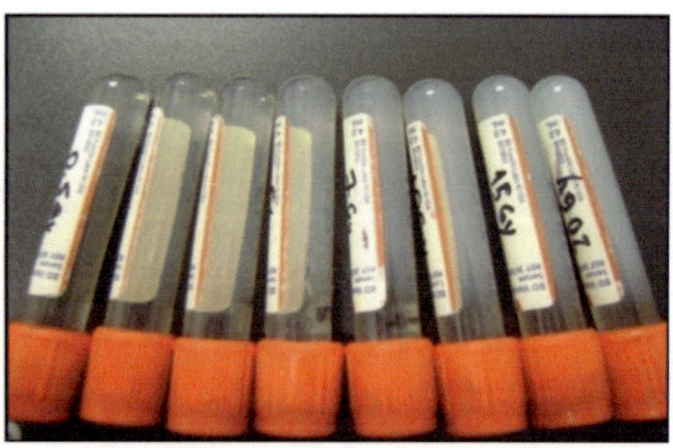

Figure 3. MAGIC-*f* irradiated with different doses.

The results obtained from the evaluation of Magic-*f* gel with the variation of the dose and dose rate are shown in figure 4.

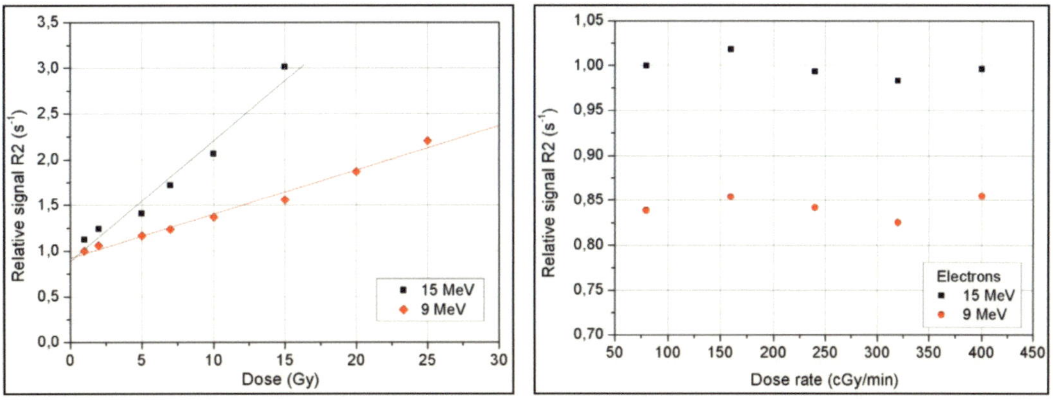

(a) (b)

Figure 4. MAGIC-f response to energies of 9 and 15 MeV: (a) variation in dose and, (b) variation in dose rate.

For all measurements the maximum uncertainty of 1.8% was found, from signal average of each irradiated homogeneous region. This was calculated through the mean of three acquisition images for each measurement.

From figure 4 (a) it can be observed that Magic-*f* gel show a dependence to different energies, with a high variation of 50% when the signal R2 from both energies to the absorbed dose of 15 Gy is compared. The linearity of the curves show correlation coefficient, r2, 0,9819 e 0,9916 for energies of 9 and 15 MeV, respectively.

The curve of the rate dose-response of the gel, shown in figure 4 (b), and the linearity curves the in figure 4 (a) show the dose dependence of the gel and maximum variations of 1.7% and 3.4% were found for energies of 9 and 15 MeV, respectively.

Figure 5 shows the phantoms irradiated for determination of PDD curves. PDD curves obtained with the gel are shown in figure 6, which were compared with the PDD obtained through ionization chamber (ic).

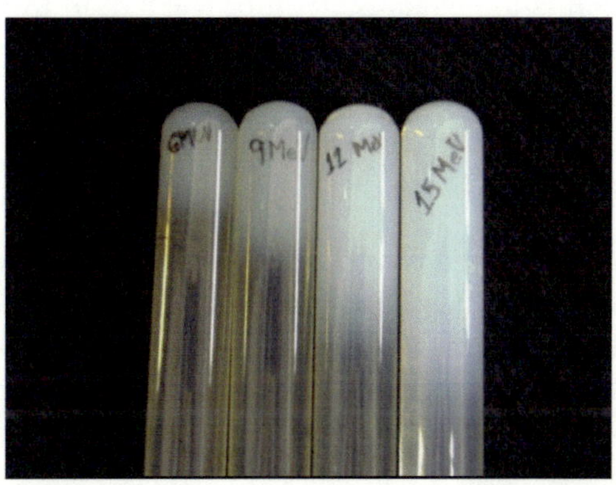

Figure 5. MAGIC-*f* irradiated with different energies.

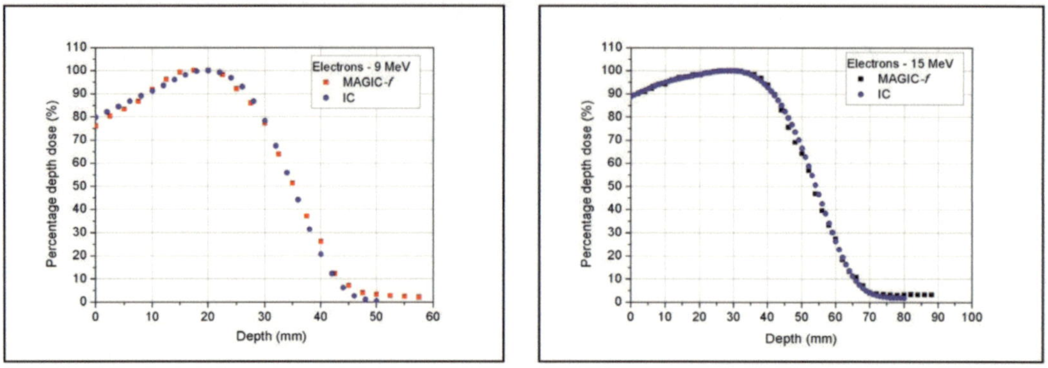

(a) (b)

Figure 6. PDD obtained with Magic-*f* gel and ionization chamber for two energies: (a) 9 MeV e (b) 15 MeV.

The maximum percentage different of 4.0% was found on comparison of PDD curves obtained with the Magic -*f* gel and ionization chamber for energies of 9 and 15 MeV.

5.1.3. Conclusion

From the results we can affirm that MAGIC-*f* dosimeter can be used as a complementary dosimetric tool for determination of the characteristics of the clinical electrons beams.

5.2. Mixed dose distribution of electron and photon beams through the gel dosimeter MAGIC-f and PENELOPE-Monte Carlo Simulation

Combining electron and photon fields in the same radiation plan can improve dose distributions, delivering a homogeneous dose to the target while reducing the dose to normal tissues. This treatment technique can benefit from both the finite range of the electrons and the sharper penumbra of the photons.

The aim of this application is to evaluate the improvement in the dose distributions from treatments using mixed photon and electron beams through polymer gel dosimetry with MAGIC-*f* and Monte Carlo simulation using PENELOPE.

5.2.1. Materials and methods

A cylindrical phantom with dimensions of 10 cm diameter and 12 cm height was homogeneously filled with MAGIC-*f*. The phantom was irradiated with a 6 MV photon beam and a 12 MeV electron beam. Field sizes of 3 x 7 cm^2 at 100 cm SSD were used to deliver a prescribed dose of 8 Gy for each beam. The phantom analysis followed a previous developed protocol in which an MRI image is registered one day after irradiation. A 3.0 T MRI scanner using a head coil and a multiple spin echo sequence with 16 echos, TE = 22.5 ms and TR = 3000 ms was used for readings. From the MRI images, R2 values were calculated on a pixel-by-pixel basis to produce R2 maps related the absorbed dose. The same geometry used in the irradiation process was simulated by PENELOPE with spatial resolution of 1 mm. The depth doses and dose profiles were used to compare the results from experiments (MAGIC-*f*) and simulation.

5.2.2. Results and discussions

The dose distributions obtained with Monte Carlo simulation are presented in the figure 7 and the dosimetric parameters obtained with PENELOPE and MAGIC-*f* are presented in figure 8.

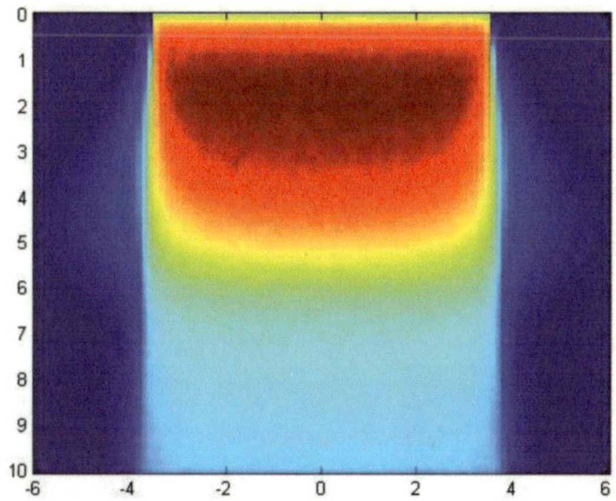

Figure 7. Dose distribution obtained with PENELOPE.

The comparisons between PENELOPE and MAGIC-*f* showed maximum differences of 3.0% and 3.2%, inside the volume of the 90% isodose for the beam profile and for the PDP curves, respectively.

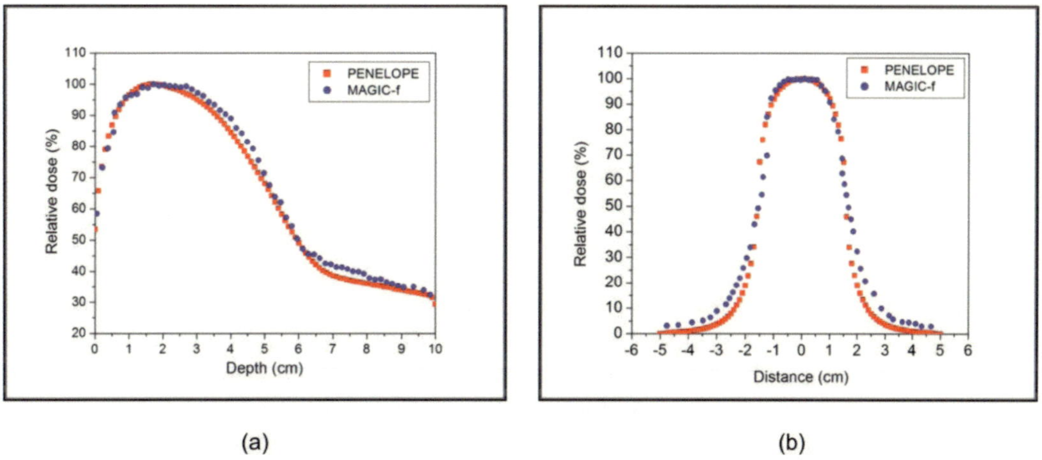

Figure 8. Dosimetric parameters obtained with PENELOPE and MAGIC-f: (a) PDD; (b) beam profile.

5.2.3. Conclusions

The comparison between the dose distribution for PENELOPE and MAGIC-*f* showed that gel dosimeter can be used in radiotherapy, for special applications, as photons and electrons mixed fields. Also, the results showed that the mixed fields reduce the absorbed dose in entrance of the prescribed field, compared with the typical electron treatment.

5.3. Evaluation of collimated fields with electron beam through XiO treatment planning system and PENELOPE Monte Carlo simulation

The determination of dose distribution by system of planning and simulation codes is different mainly due to calculation algorithm. Dose distribution may vary depending upon the dosimetric parameters, for example, the field size. The PDP for collimated fields were evaluated by the XiO treatment planning system (TPS) and PENELOPE Monte Carlo simulation. Figure 9 shows the dose distribution obtained for different field size for the 9 MeV.

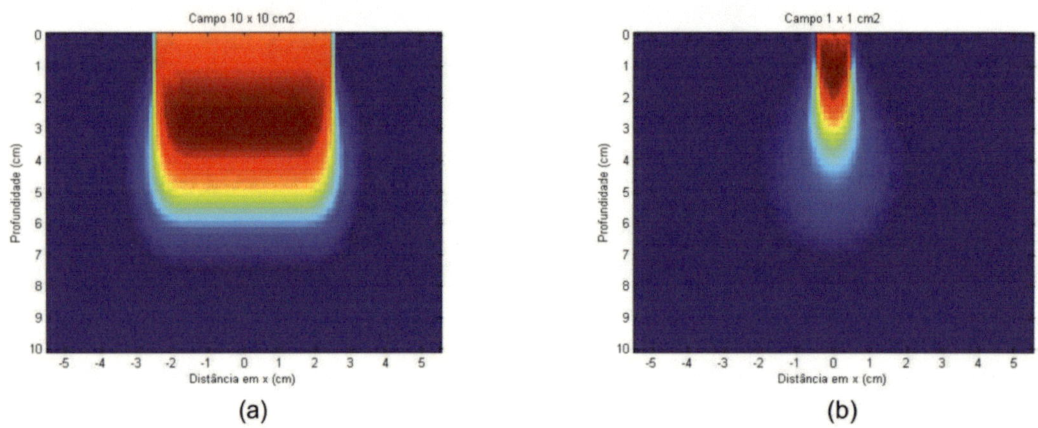

Figure 9. Dose distribution obtained for different field size for the 9: (a) 10 x 10 cm^2, (b) 1 x 1 cm^2.

5.3.1. Materials and methods

Using the standard applicators for electrons beam of 10x10 cm^2 beam profiles through PENELOPE, TPS and ionization chamber (0.1cc/IBA) were determined. From the concordances between the two calculation algorithm, simulation code and TPS, were studied for collimated fields. The standard applicator of 10 x 10 cm^2 and blocks of cerrobend were used to collimate fields of 1x1 , 3x3 and 5x5 cm^2, for 9 MeV beam(fig:10). The PDD obtained by the code and TPS were analyzed with the MatLab® software.

Figure 10. Cerrobend collimator.

5.3.2. Results and discussions

A maximum difference of 1.5 % when comparing the values obtained from the PENELOPE and ionization chamber for PDD obtained at the maximum depth dose and 2.2 % when TPS and ionization values were compared, for the two applicators respectively. A maximum difference of 3.0% and 3.2% were also found on comparing with other depth using PENELOPE and TPS. These differences increase to 5% for isodose less than 50 %, as shown in figure 11.

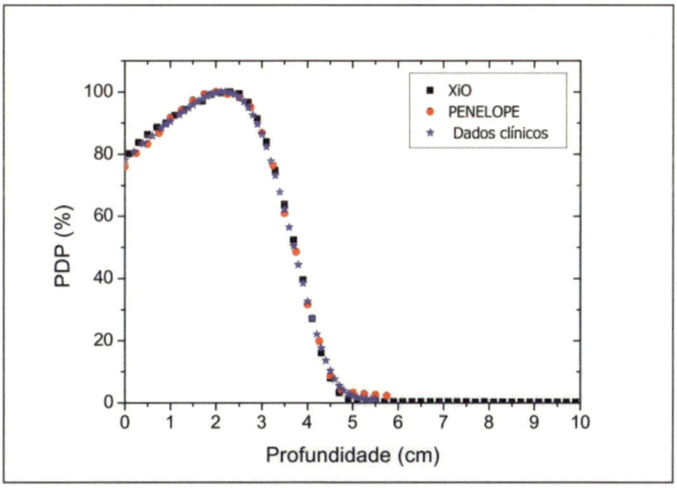

Figure 11. PDD in reference condition for dosimetric tool: XiO, PENELOPE and ionization chamber

The comparison of the PDD obtained at depth greater than 50% showed maximum difference of 5.0 %, 4.3 %, 4.8 %, respectively for each studied field. These differences increase to 12% for other depth, as shown in figure 12.

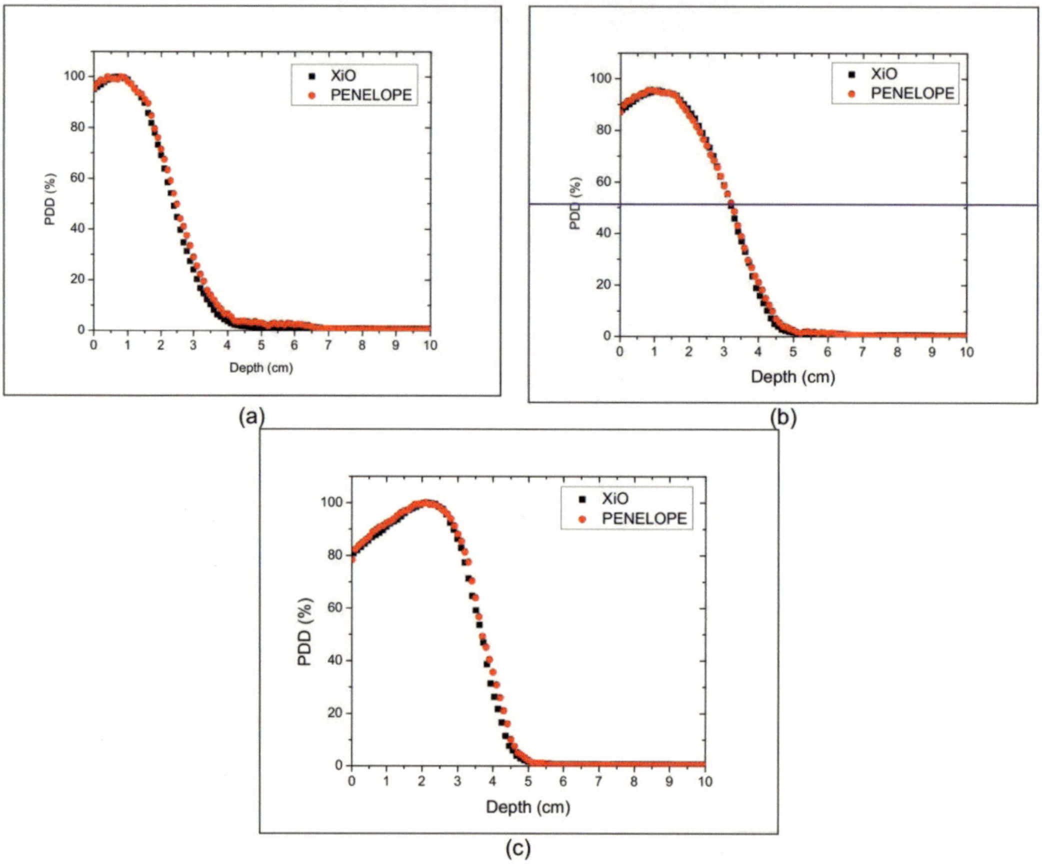

Figure 12. PDD obtained for 9 MeV in different fields size: (a) 1 x 1 cm², (b) 3 x 3 cm², (a) 5 x 5 cm².

5.3.3. Conclusion

The TPS curves did not show a continuous behavior due the interpolation of data for these collimated fields. From the results it can be inferred that despite the differences of both calculation algorithm the behavior of the beam profiles was similar.

5.4. Study of different materials for a conformational simulation in radiotherapy using the PENELOPE-Monte Carlo code

The use of conformal techniques for photons beam represent the most modern procedures in radiotherapy, like intensity modulation radiation (IMRT), the intra-operative radiotherapy (IORT) and tomotherapy. For photon beams, irregular fields are obtained through shielding blocks of high atomic number, specially manufactured for each patient, or by liear accelerator accessories such as multi-leaf collimators. Since this collimation enables better targeting of treatment of the target volume while protecting surrounding healthy tissues, the use of conformal techniques in radiotherapy with photons beams has made it possible to increase the prescribed dose compared with those used in conventional techniques.

For electron beams, currently, the irradiation units are not fitted with suitable accessories to give a conformal technique, although with the technological progress of the radiotherapy and the improvement of the algorithms used in the treatment planning system, radiotherapy with modulated electron (MERT) beams may be a more tangible possibility. Thus, the dosimetric characteristics of the irradiation fields produced by this proposal have been investigated in this study.

Recent authors have studied different possibilities for realization of this new radiotherapy technique, An example being the construction of the multi-leaf collimators for a specific electron beam [32]. The possibility of using the multi-leaf collimators using only photon beams[33] or development of collimators additional to maintain a standard feature of the treatment with electrons beams [34,35].

However, for the modality of MERT the major limitation is the thickness of the additional collimators used because of the short distance between the applicator and irradiated surface pattern, requiring investigation of the possibility of using high atomic number materials in the manufacture of additional collimator.

Hence the additional optimization of collimators may be determined using computational simulation, which is a useful alternative to the experimental methods, it has the advantaged of providing detailed studies and in different experimental conditions without using methodologies that are time-consuming and costly [36].

The proposition of this study is to analyze using Monte Carlo simulation with the PENELOPE code to determination of dose distribution, PDD, and dose profiles obtained with the MERT technique with additional collimators of different material: cerrobend (cerr) and acrylic (PMMA)

5.4.1. Materials and methods

The different dosimetric response of collimator for the treatment of MERT were evaluated using Monte Carlo simulation with PENELOPE code, version 2008.The geometry of simulation is shown in figure 13.

In this study, we used spectra electron beam 6 and 15 MeV specific for the linear accelerator Clinac 2100 C (Varian) irradiating an object simulator of 20 x 20 x 20 cm^3 filled with water. The SSD used was 100 cm, the irradiation field of 10 x 10 cm^2, and collimated by the additional applicator for an irradiation field of 1 x 1 cm^2. PDD and the beam profile in the depth of treatment (85% isodose) were determined by PENELOPE code for both materials and energy, with spatial resolution of 1 mm along the central axis of the radiation field. Were also determined the distribution of doses deposited in planes parallel and perpendicular to the central axis of the radiation beams used.

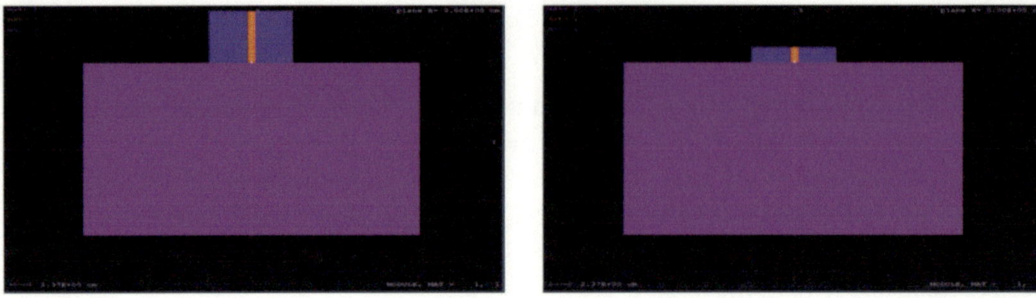

(a) (b)

Figure 13. Geometry simulated for spectrum of 6 MeV with different collimators additional: (a) PMMA and (b) Cerrobend.

The thickness of the collimators additional cerrobend and PMMA were determined using the same attenuation in different materials and different energies. Table 1 shows the thicknesses used in the simulation.

Energy (MeV)	Thickness of the additional collimator (cm)	
	Cerr	PMMA
6	1,8	6,0
15	3,3	10,9

Table 1. Thickness for both additional collimators

5.4.2. Results and discussions

The doses distribution of one plane is represented in phantom is shown in Figure 14, showing, qualitatively, the difference in dose distribution obtained for the same collimator additional material, acrylic, for both radiation beams 6 and 15MeV.

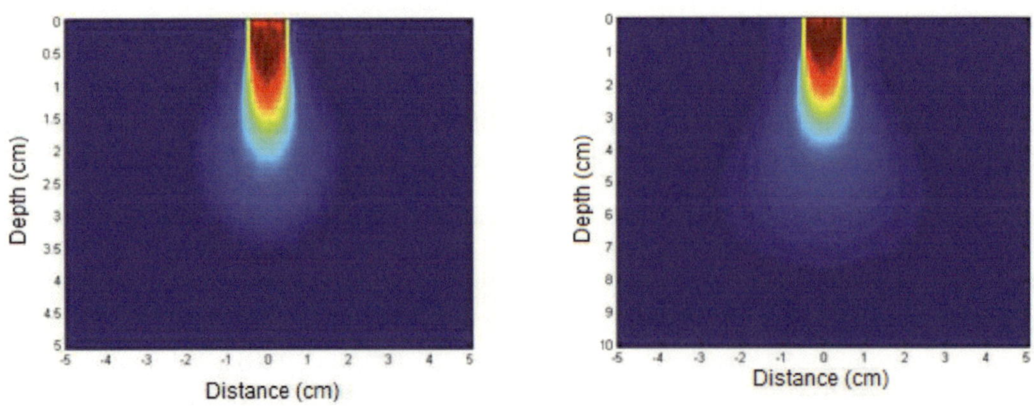

(a) (b)

Figure 14. Dose distributions with the acrylic collimator for the energies of: (a) 6 MeV, (b) 15 MeV.

The obtained dosimetric responses for different material and energies are presented in figures 15 and 16.

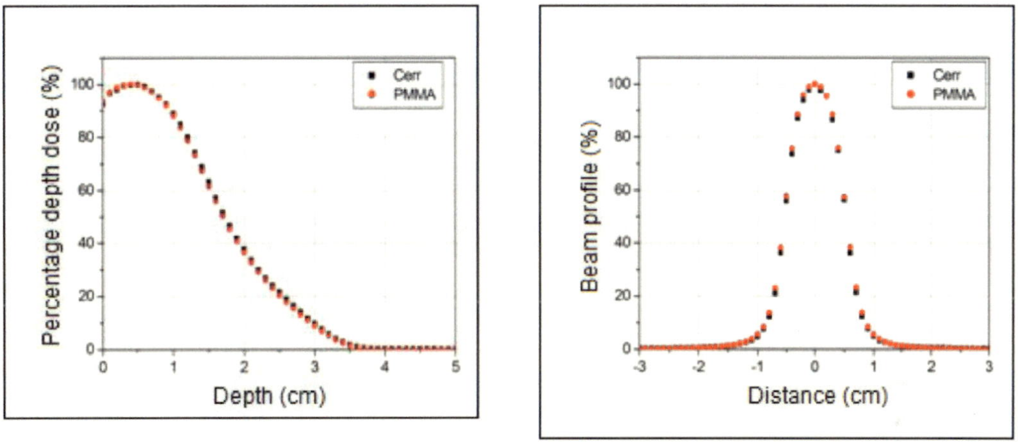

(a) (b)

Figure 15. Comparison of the two results obtained with the Cerrobend and PMMA for energies of 6 MeV: (a) PDD, (b) Beam profile.

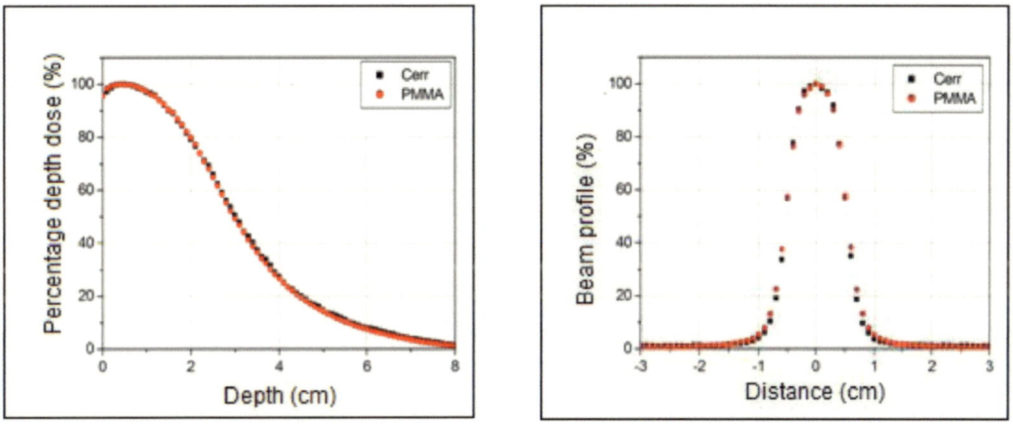

(a) (b)

Figure 16. Comparison of the two results obtained with the Cerrobend and PMMA for energies of 15 MeV: (a) PDD, (b) Beam profile.

Table 2 shows, quantitatively, the major differences from dosimetric parameters, PDD and beam profile at treatment depth, with collimators cerrobend and PMMA for energies of 6 MeV and 15 MeV

Since the irradiation characteristics of an electron beam, it was expected that there were greater photon contamination when the collimator was added a material of high atomic number such as cerrobend, however, results presented in Figures 15 and 16 show that the thickness of the collimator is added to attenuate photons produced contamination is reduced, with the same responses observed with the material suitable for collimator of electrons, such as PMMA, with the advantage of lower thickness.

Energy (MeV)	PDD (%)	Beam Profile (%)
6	1,2	2,5
15	1,5	3,2

Table 2. Percentages of the major differences for the dosimetric parameters, were evaluated the additional collimators (cerr and PMMA) in the energies of 6 MeV and 15 MeV.

5.4.3. Conclusions

It can be inferred therefore, that the additional collimator for the proposed technique can be manufactured using a material of high atomic number, conserving dosimetric characteristics already established, with the advantage of lower thickness.

Author details

Thatiane Alves Pianoschi and Mirko Salomón Alva-Sánchez
Department of Physics, University of São Paulo, Ribeirão Preto, SP, Brazil

Acknowledgement

The presented works were supported by: CAPES (Coordination of Improvement of Higher Education Personnel- *Portuguese: Coordenação de Aperfeiçoamento de Pessoal de Nível Superior*) and the physics department of the University of Sao Paulo-Campus Ribeirao Preto-Brazil.

We are thankfully to the personnel from the Cancer Hospital of Barretos, especially to Marcelo Carvalho Santanna and Fransico Américo Marcelino. Hospital of Clinics from Ribeirão Preto, Hospital Sírio Libanês- Sao Paulo-Brazil, especially to Cecilia Hadad.

The authors are thankfully to the professor Ph. D. Patricia Nicolucci to orientation for all the works, which are part of our doctorate work.

We are thankfully for technical support to José Luiz Aziani and Carlos Renato da Silva is also appreciated.

6. References

[1] Hogstrom K, Almond P (2006) Review of electron beam therapy physics. Phys. Med. Biol. 51 R455–R489.

[2] Khan F, Doppke K, Hogstrom K (1991) Clinical electron-beam dosimetry: Report of AAPM Radiation Therapy Committee Task Group No. 25. Med. Phys. 18: 73-107.

[3] Sempau J, Fernandez-Varea J, Acosta E (2003) Experimental benchmarks of the Monte Carlo code PENELOPE. Nuc. Instrum. Methods Phys. Res. B 207:107-123.

[4] Rogers D (2006) Reviews: Fifty years of Monte Carlo simulations for medical physics, Phys. Med. Biol. 51:R287 – R301.

[5] Day M, Stein G (1950) Chemical effects of ionizing radiation in some gels. Nature. 166:141-147.

[6] Andrews H, Murphy R, LeBrun E (1957) Gel dosimeter for depth dose measurements. Rev. Sci. Instrum. 28:329-332.

[7] Gore J, Kang Y, Schulz R (1984) Measurement of radiation dose distributions by nuclear magnetic resonance (NMR) imaging. Phys. Med. Biol. 29:1189-1197.

[8] Maryanski M, Gore J, Kennan R, Schulz R (1993) NMR relaxation enhancement in gels polymerized and cross-linked by ionizing radiation: a new approach to 3D dosimeter by MRI. Mag. reson. imaging. 11: 253-258.

[9] Maryanski M, Schultz R, Ibbott G, Gatenby J, Xie J, Horton D, Gore J (1994) Magnetic resonance imaging of radiation dose distributions using a polymer-gel dosimeter. Phys. med. biol. 39: 1437-1455.

[10] Baldock C, Burford R, Billinghan N, Wagner G, Patval S, Badawi R, Keevil S (1998) Experimental procedure for the manufacture of polycrylamide gel (PAG) for magnetic resonance imaging (MRI) radiation dosimetry. Phys. med. biol. 43: 695-702.

[11] Fong P, Keil D, Does M, Gore J (2001) Polymer gels for magnetic resonance imaging of radiation dose distributions at normal room atmosphere. Phys. med. biol. 46: 3105-3113.

[12] Baldock C, Burford R, Billinghan N, Wagner G, Patval S, Badawi R, Keevil S (1998) Experimental procedure for the manufacture of polycrylamide gel (PAG) for magnetic resonance imaging (MRI) radiation dosimetry. Phys. med. biol. 43: 695-702.

[13] De Deene Y, Hurley C, Venning A, Vergote K, Mather M, Healy B, Baldock C (2002) A basic study of some normoxic polymer gel dosimeter. Phys. med. biol. 47: 3441-3463.

[14] Gustavsson H, Karlsson A, Back S, Olsson L, Haraldsson P, Engstrom P, Nystrom H (2003) MAGIC-type gel for three-dimensional dosimetry: Intensity-modulated radiation therapy verification. Med. phys. 30:1264-1271.

[15] Fernandes J, Pastorello B, de Araujo D, Baffa O (2008) Formaldehyde increases magic gel dosimeter melting point and sensitivity. Phys. med. biol. 53: N1-N6.

[16] Kramer G, Crowley P, Burns L (2000) Investigation the impossible: Monte Carlos simulation Radiat. Prot. Dosimetry. 89: 259 – 262.

[17] Bielajew A, Rogers D (1992) A standard timing benchmark for EGS4 Monte Carlo calculation Med. Phys. 19:303 – 304.

[18] Hendrikcs J, Adams K, Booth T (2001) Present and future capabilities of MCNP. App. Radiat. Isotopes. 53: 857 – 861.

[19] Salvat F, Fernández-Varea J, Sempau J (2008) A Code System for Monte-Carlo Simulation of Electron and Photon Transport France.

[20] Allison J, Amako J, Apostolakis H (2006) Geant4 developments and applications. IEEE Transactions on nuclear science. 53: 70-78.

[21] Brualla L, Palanco-Zamora R, Witting A (2009) Comparison between PENELOPE and electron Monte Carlo simulations of electron fields used in the treatment of conjunctival lymphoma Phys. Med. Biol. 54:5469-5481.

[22] Duclous R, Dubroca B, Frank M (2010) Deterministic partial differential equation model for dose calculation in electron radiotherapy. Phys. Med. Biol. 55: 3843-3857.

[23] Casado F, Garcia-Pareja S, Cenizo E (2010) Dosimetric characterization of an Ir-192 brachytherapy source with the Monte Carlo code PENELOPE. Physica Medica. 26:132-139.

[24] Sempau J, Badal A, Brualla L (2011) A PENELOPE-based system for the automated Monte Carlo simulation of clinacs and voxelized geometries—application to far-from-axis fields. Med. phys. 38: 5887-5895.

[25] Koivunoroa H, Siiskonen T, Kotiluoto P, Auterinen I, Hippelainen E, Savolainen S (2012) Accuracy of the electron transport in MCNP5 and its suitability for ionization chamber response simulations: A comparison with the EGSNRC and PENELOPE codes. Med. phys. 39:1335-1344.

[26] Ramirez J, Chen F, Nicolucci P, Baffa O (2011) Dosimetry of small radiation field in inhomogeneous medium using alanine/EPR minidosimeters and PENELOPE Monte Carlo simulation. Radiat. Measur.46:941-944.

[27] Gonzales D, Requena S, Williams S (2012) Au La x-rays induced by photons from 241Am: Comparison of experimental results and the predictions of PENELOPE. Appl. Radiat. Isot. 70: 301–304.

[28] International commission on radiation units and measurements, ICRU, (1989) Prescribing, Tissue substitutes in radiation dosimetry and measurement, ICRU Report 44, EUA.

[29] Hendee W, Ibbott G, Hendee E (2005) Radiation Therapy Physics. Wiley-Liss Publ. ISBN 0-471-39493-9.

[30] Lahanas M,Baltas D,Giannouli S (2003) Global convergence analysis of fast multiobjective gradient-based dose optimization algorithms for high-dose-rate brachytherapy. Phys. Med. Biol. 48:599-617.

[31] Galvin M, Ezzel G, Eisbrauch A, Yu C, Butler B, Xiao Y, Rosen I, Rosenman I (2004) Implementing IMRT in clinical practice: a joint document of the American Society for Therapeutic Radiology and Oncology and the American Association of Physicists in Medicine. Int. J. Radiat. Oncol. Biol. Phys. 58: 1616–34.

[32] Gauer T, Albers D, Cremers F, Harmansa R (2006) Design of a computer-controlled multileaf collimator for advanced electron radiotherapy. Phys. Med. Biol. 51: 5987-6003.

[33] Jin L, Ma M-C, Fan J (2008) Dosimetric verification of modulated electron radiotherapy delivered using a photon multileaf collimator for intact breasts. Phys. Med. Biol. 53: 6009-6025.

[34] Al-Yahya K, Verhaegen F, Seuntjens J (2007) Design and dosimetry of a few leaf electron collimators for energy modulated electron therapy. Med. Phys. 34(12): 4782-4791.

[35] Vatanen T, Traneus E, Lahtinen T (2008) Dosimetric verification of a Monte Carlo electron beam model for na add-on eMLC. Phys. Med. Biol. 53(2): 391-404.

[36] Sempau J, Acosta E, Baró J (1997) An algorithm for Monte Carlo simulation of coupled electron-photon transport, Nucl. Instr. and Meth. B. 132: 377-390.

Chapter 4

3D Dosimetric Tools in Radiotherapy for Photon Beams

Mirko Salomón Alva-Sánchez and Thatiane Alves Pianoschi

Additional information is available at the end of the chapter

http://dx.doi.org/10.5772/56348

1. Introduction

The technological advances to develop or improve devices for radiotherapy have increased in recent years, leading to conformality in the absorbed-dose within the tumor volume and avoiding healthy tissue. For quality assurance of the treatment, measurement of complex dose distribution from the new devices needs to be verified in the treated volume and healthy structure surrounding it.

According to ICRU the prescribed and administered dose must be within -5 % and + 7 % [1]. Many dosimetric protocols for radiotherapy recommend use of an ionization chamber like reference dosimeter, however, dosimeters like TLD, film, diode and other tools can be used although they measure the absorbed-dose at a point or in two dimensions [2,3].

A clinical system-dosimetry that can measure the dose distribution in 3D with tissue-equivalent characteristics is not available. In this context, dosimetry based on gel has shown a great potential for measured dose-distribution in 3D [4-7].

The MAGIC dosimeter with formaldehyde has characteristics of tissue-equivalence to water, stability of response and high detection resolution for suitable application in 3D dosimetry for complex treatments. Especially for use in techniques of high dose rate it can verify the volume dose as a quality control in radiotherapy. The gel dosimeter with formaldehyde (MAGIC-*f*) has been used in simulation studies as well as in clinical [8-11].

The most common used simulation algorithms, based on Monte Carlo calculations, are ITS2, EGS4, MCNP4 and PENELOPE [12-16]. The algorithm for mixed transport of charged particles implemented by PENELOPE algorithm led to its use in radiotherapy, simulating several irradiations and geometry of clinical situations in radiotherapy, which have shown a good concordance with experimental measurements using different irradiation techniques [17-20].

This chapter will approach two dosimetry tools: the MAGIC-*f* gel dosimeter and the PENELOPE-Monte code. Also, this chapter will cover some clinical applications of the two dosimetry tools with modalities like brachytheray, radiosurgery and 3D conformal therapy.

2. MAGIC-*f* dosimeter

Gel-based dosimeters called polymer gel dosimeters have been proposed since 1954 [21-24]. In 1992, formulations like BANANA, BANG and PAG polymer gels were proposed [25,26]. The suppression of polymerization by oxygen, present in these gels, was improved with a proposition of MAGIC gel (Methacrylic and Ascorbic acid in Gelatin Initiated by Copper) [27], which allows the MAGIC polymeric gels to be prepared in normal weather conditions [28-30], which for higher temperatures beyond 25 °C loss stability. Thus, was add formaldehyde to the original formulation of the MAGIC, re-calling as MAGIC-*f*, which led to a rise in its melting point to 69 °C [31].

The monomeric compounds of the MAGIC-*f* dosimeter immersed in an aqueous matrix, after exposure to irradiation suffer a polymer reaction, resulting in a polymer gel matrix. This formation changes the NMR relaxation properties of the gel and can be related to the deposited energy. One way to show this change in 3D is by using magnetic resonance images.

3. PENELOPE code

PENELOPE (Penetration and Energy Loss of Electrons and Positron) is a package used to simulate the transport of electrons, positrons and photons in arbitrary materials and complex geometries. The package has written on a FORTRAN platform, which include characteristics of many materials, and a user file. Besides these files, the code has a database with the characteristics of various materials of interest in radiological physics [32], cross section libraries and other quantities necessary for the transport of particles [33]. The simulation algorithm is based on a model that combines numerical and analytical cross sections for different types of interactions and is applied to initial energies from 1 keV to 1 GeV. The code has been used in some applications to simulate different irradiation techniques, reproduce the dimensions of irradiation geometry, source, energy, distance and other dosimetric parameters.

4. Application of the two dosimetric tools for some modalities in radiotherapy

4.1. Water-equivalent calibration of ^{192}Ir HDR brachytheray source using MAGIC-*f* gel

HDR brachytherapy is a treatment technique that uses sealed radioactive sources to treat prostate, colorectal cancer and some gynecological malignancies [34,35]. The commonly used implants are of ^{192}Ir sources with a half-life of 74 days, emitting beta rays ranging from 530 keV to 670 keV, and a gamma ray with an energy of 370 keV [36].

Due to the energy spectrum present in this source and the high gradient dose, a proper source calibration is necessary, although vendors assign large uncertainties to the calibration values (up to ± 10%). The calibration protocols for HDR brachytherapy sources recommend a specification in reference to air kerma rate using ionization chamber [37,38]. Absorbed dose measurement in water has recently been proposed to calibrate ^{192}Ir HDR brachytherapy sources [39].

For determination of the water-equivalent calibration of an ^{192}Ir HDR source we used the MAGI-f dosimeter, which has been shown to be a suitable dosimeter for many treatments in radiotherapy due to its characteristics of water equivalence (effective atomic number of 7.41), spatial resolution better than 1mm and the most important characteristic of measuring dose tridimensionally. Through the MAGIC-f were determined dosimetric parameters like percentage dose depth (PDD), and calibration curve (CC).

The two dosimetric parameters, PDD and CC, were obtained with the three dosimetric tools a)thermoluminescent dosimeter, b)ionization chamber and c)PENELOPE-Monte Carlo code simulation. The results obtained with these dosimetric tools were compared with the obtained data of polymer gel [40].

4.1.1. Materials and Methods

A water phantom of 50 x 50 x 50 cm^3 with electronic positioning holder (0.1mm precision) and the Gamma Med Plus 232 ^{192}Ir source with 5.43 Ci (200.91 GBq) was employed in this work. This source has an active cylindrical volume of 0.6 mm diameter and 3.5 mm height encapsulated in stainless steel welded to a steel cable, as shown in figure 1.

Measurements for PDD determinations, in water, were carried out using TLDs (LiF-100 with dimensions: 0.9 x 0.9 x 3.1 mm^3) packed with plastic and a plane parallel ionization chamber (Markus type, 0.05cm^3) after proper calibration. Both dosimeters were introduced in different depths (from 3 up to 12 mm) in a water phantom (figure 2). The catheter containing the HDR source was set parallel to the TLDs and the ionization chamber.

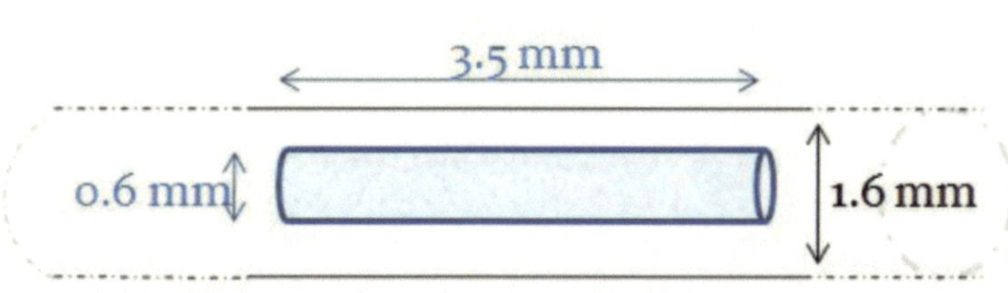

Figure 1. Design of ^{192}Ir source

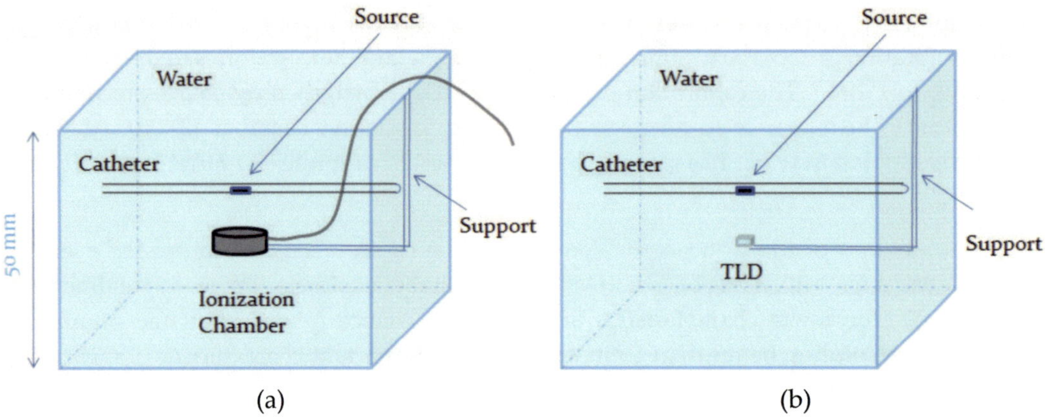

Figure 2. Measurements setup for PDD determination with (a) ionization chamber and (b) TLDs.

Magic gel procedures

After preparation the solution of gel, this was poured in an acrylic cylindrical phantom of 6 cm diameter, 8 cm height and 0.5 cm thickness taking care to prevent air bubbles(figure 3).

During the PDD measurement the phantom filled with the gel was kept inside the water phantom and the source was positioned at a distance of 2 cm from the gel phantom wall.

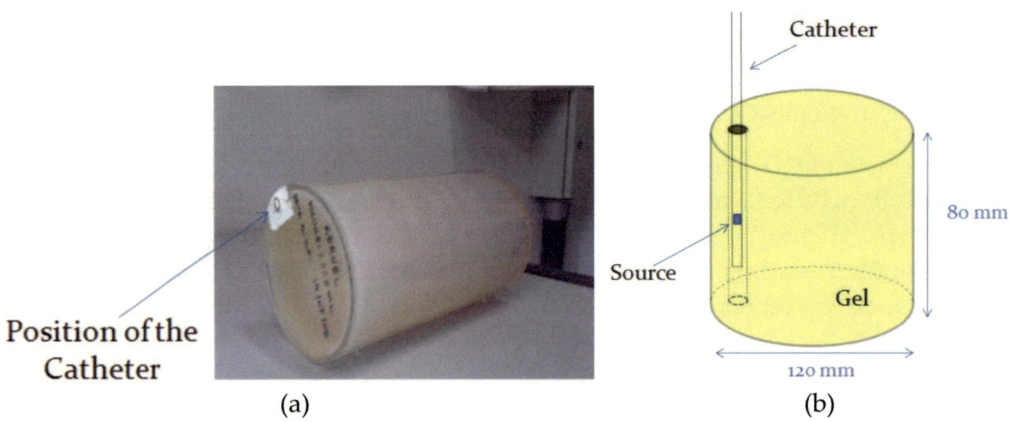

Figure 3. Cylindrical phantom to obtain the PDD curve: (a) with the MAGIC-f gel and (b) scheme of the irradiation setup.

For gel calibration, part of the gel was poured into three cylindrical glass tubes with 5ml volume, 12mm diameter and 75 mm height, closed with a 20mm hermetic stopper inserted in a plastic cover, as shown in figure 4.

Relaxation images of the gel phantom and glass tubes were acquired using a 1.5 T scanner (Siemens, Magneton Vision) one day after the irradiation to allow enough time for gel reaction completion. A head coil and multi spin echo sequences with 16 echo times in multiples of 22.5 ms, a repetition time of 3000 ms and a matrix size of 512 × 512 pixels were used during image acquisition. The slice thickness was 2mm. The transverse relaxation rate R2 (=1/T2) was calculated by fitting the signal intensities versus the echo time pixel by pixel. The R2 maps were related to absorbed dose using a specific program developed in MatLab® 6.5.

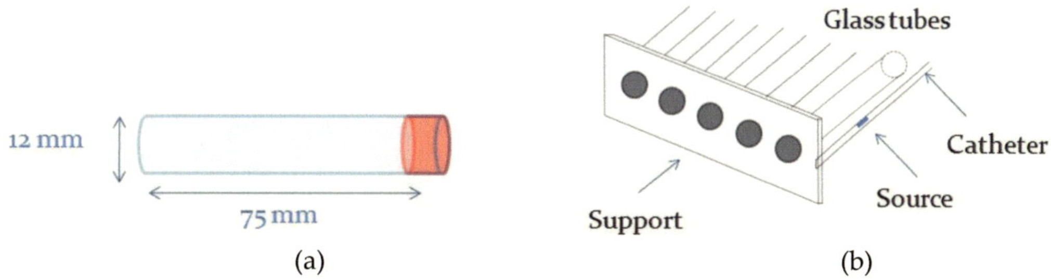

Figure 4. (a) Glass tube of 5 ml and (b) scheme of the setup to determine the calibration curve for the gel dosimeter.

PENELOPE-Monte Carlo

The user code was written to describe the geometry of the source and simulate all anisotropic properties of the brachytherapy extended source. A cubic water phantom were simulated, with dimensions of 30 cm, appropriate to obtain full scattering conditions. Simulation spatial resolution was set to1 mm.

4.1.2. Results and conclusions

Calibration curves for HDR ^{192}Ir source using TLD, ionization chamber and MAGIC gel are shown in figure 5.

The three dosimeters showed linearity with relative dose with correlation coefficients (R) of 0.9948, 0.9959 and 0.9971 for CI, TLD and the gel respectively. Maximum uncertainty of 3.5 % was found for measurements with the TLD for 10 Gy and a maximum uncertainty of 0.8 % was found for 10 Gy using the other dosimeter. The CC curve obtained with the TLDs did not show sensitivity for low doses (0.5 Gy) and MAGIC gel presented a background dose, approximately of 1.3 s^{-1}, which was subtracted for PDD measurements.

The calibration curve obtained with the Magic-f gel for the HDR ^{192}Ir source shows that it is possible to calibrate this source in water as stipulated in the protocol for external source [41]. Since, the gel is constituted by 90% water, effective-atomic-number of 7.41, which can be represent interaction of radiation with the water.

For PDD determinations, dose map distributions were obtained with MAGIC-f in a few planes as shown in figure 6.

The PDD curves were determined through percentage absorbed dose in depth, D_d, relative to the absorbed dose in the maximum depth, D_m, as show in equation 1

$$PDD(d) = \frac{D_d}{D_m} x\, 100\% \quad (1)$$

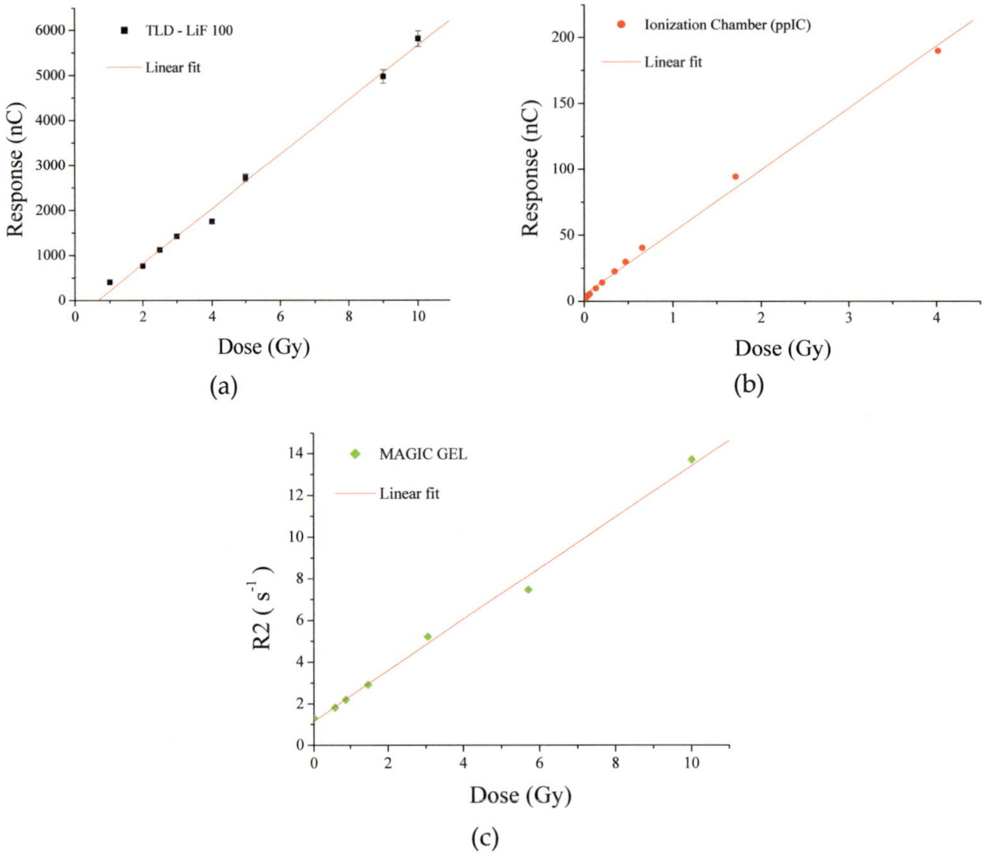

Figure 5. HDR ^{192}Ir source calibration curves: (a) TLD LiF-100 , (b) IC and (c) MAGIC polymer gel

The dose maps were obtained through the relaxation images and analysed with the program, in MatLab® developed to relate the nuclear resonance signal of the image with the absorbed dose.

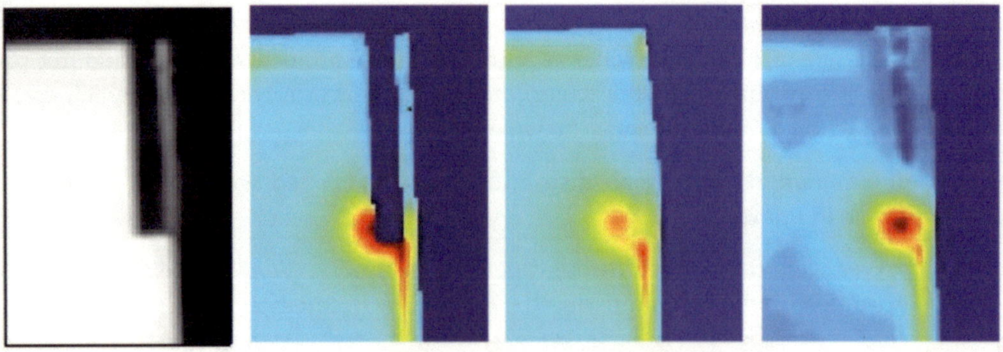

Figure 6. Relaxometry image and dose maps around the HDR ^{192}Ir source obtained experimentally with MAGIC polymer gel.

The PDD curves obtained by TLD, ionization chamber, MAGIC gel and PENELOPE code are shown in figure 7.

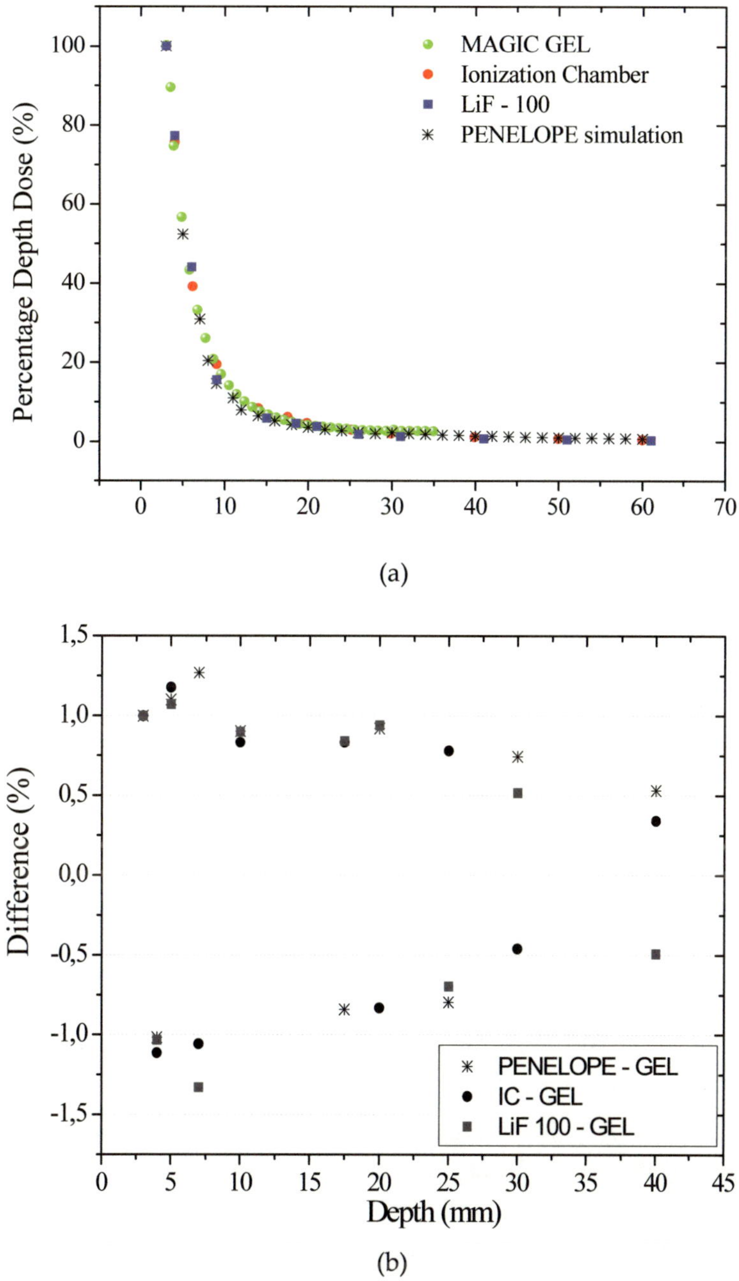

Figure 7. Percentage depth dose for HDR ^{192}Ir with MAGIC polymer gel, ionization chamber, TLD LiF-100 and PENELOPE code, (b) differences between PENELOPE, IC and TLD-LiFs related to MAGIC gel.

From the percentage depth dose and calibration curve results it can be concluded that the MAGIC polymeric gel can be used as dosimeter in routinely clinical procedures and can be used to check the calibration of ^{192}I sources in an water-equivalent medium as part of the assurance quality in HDR brachytherapy.

4.2. Dose-distribution of small fields through the MAGIC-*f* dosimeter and PENELOPE code

Some techniques in radiotherapy, like intensity-modulated radiation and stereotactic radiosurgery, use small radiation fields and higher dose-radiation. Since the dosimetry of the small fields is more critical than that for large fields, due to no-equilibrium conditions created as consequence of the secondary electron track length and the real source size project from the collimator to the phantom or patient. Other complication is the perturbation of the level of disequilibrium, which is inducted bye the dosimeter when introduced in the track of the radiation fields. Thus, due these characteristics of the small fields, an accurate dose verification of the delivery dose of the small radiation fields is required [42-45].

The aim of this application is evaluate the response of MAGIC-f dosimeter, after being exposed to a conformal irradiation technique using a small field to compare the results with PENELOPE code through two dimensional dosimetric parameters [46].

4.2.1. Materials and Methods

Treatment planning system (TPS)

The TPS BrainSCAN 5.31, was used to plan a conformal irradiation with the tomography images of the cylindrical phantom, with 9 irradiation fields from 0° to 340°. The prescribed dose was determined through the calculated algorithm using pencilbeam for 16 Gy in 100% of the target volume.

MAGIC-f gel and PENELOPE code

The solution of Magic-f gel was poured in a cylindrical phantom of PPMA with dimensions of 12 cm diameter, 18 cm height and thickness 0.5 cm. The phantom containing the gel was placed at source-phantom center distance of 100 cm, figure 8, from the linear accelerator (Siemens MEVATROM with multileaf MX2), irradiated with 9 fields of 6 MV photon beam, each one of 1 x 1 cm^2, to a total dose of 16 Gy. The experimental irradiation was made in the same irradiation conditions as the TPS.

PENELOPE simulations were used to determine the percentage depth dose, beam profile and a conformational dose distribution in the same irradiation and geometry condition made by the MAGIC-f.

After the irradiation, the reading was performed through MRI using a 3 T scanner and head coil. Relaxation images were obtained using a multi spin echo sequence with 16 echo times in multiples of 22.5 ms, a repetition time of 3000 ms and a matrix size of 256 × 256 pixels. The slice thickness was 2mm for each slice.

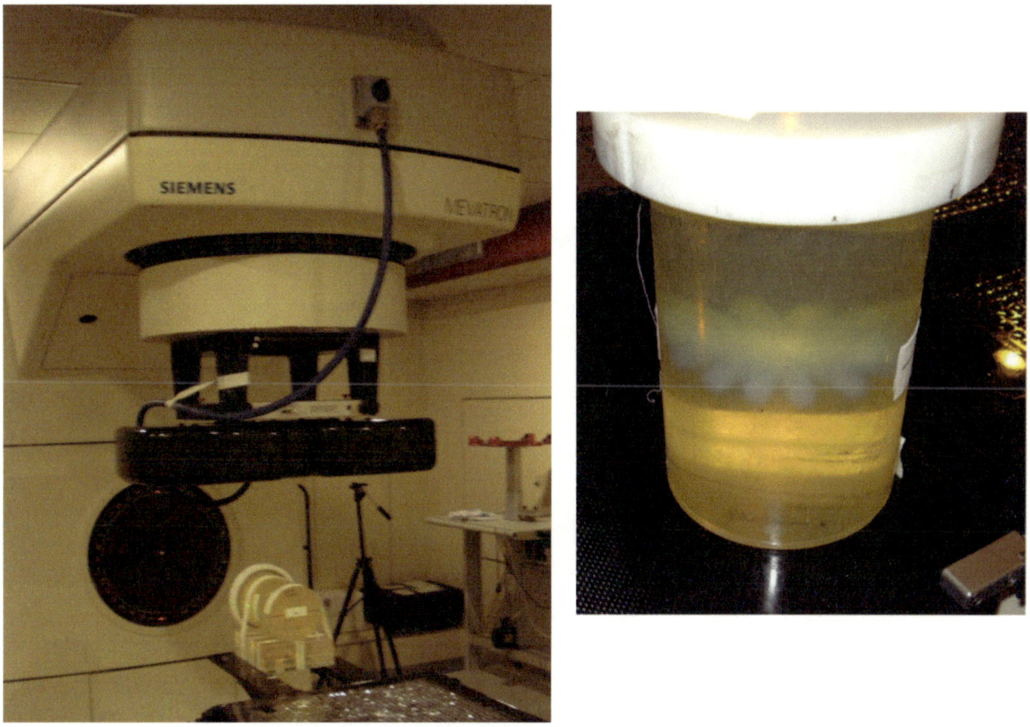

(a) (b)

Figure 8. (a) Setup of the irradiation with the linear accelerator coupled with the multileaf, (b) cylindrical phantom after irradiation with 9 fields

PENELOPE simulations were used to match the irradiation and geometry conditions as the experimental irradiation. The spatial resolution for the simulations was 0.5 mm.

The two dose distributions obtained with the gel and PENELOPE were compared through beam profiles for each irradiated field.

4.2.2. Results and conclusions

The two dosimetric tools were validated through comparison PDD and BP curves. The determination of the PDD was made through the equation 1. The PDD curves are shown in the figure 9 for an irradiation field of 1 x 1 cm^2, source-phantom center distance of 100 cm with energy of 6 MV and absorbed dose of 5 Gy.

The PDD curves obtained with the two dosimetric tools show the same behavior. Maximum differences of 6.5 % were found in 10.1 cm depth comparing the two curves.

The BP determinations were made in the line along of maximum dose depth (perpendicular to the beam irradiation) through the equation 2, since D_n are de dose in the line where is the maximum dose, D_m. The BP determinations were irradiated with the same irradiation of the PDDs. Both BPs curves are show in figure 10.

$$BP(\%) = \frac{D_n}{D_m} x\, 100\% \tag{2}$$

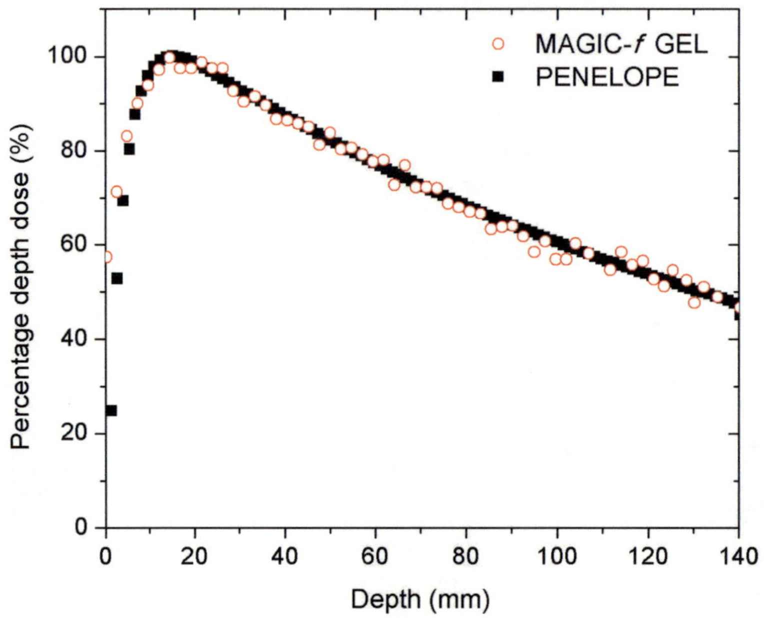

Figure 9. Percentage depth doses obtained with the MAGIC-f gel and PENELOPE code for a irradiation field of 1 x 1 cm²

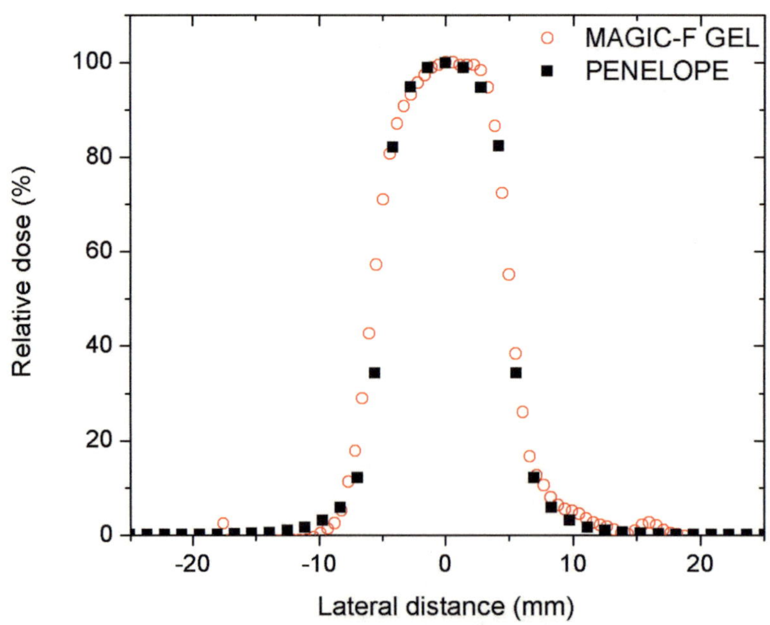

Figure 10. Beam profile obtained with the MAGIC-f gel and PENELOPE code for 1 x 1 cm².

The BP curves obtained with the two dosimetric tools show the same behavior, with maximum differences of 4.3 %, where compare both curves in distances more than size of the irradiation field. Inside the irradiation field (1 cm^2) the maximum difference was of 0.9 %.

The dose-distributions to a conformal irradiation technique obtained with tool dosimeters, gel and PENELOPE, are showed in the figure 11. The dose-distributions obtained were compared through the BP, as show in figure 12.

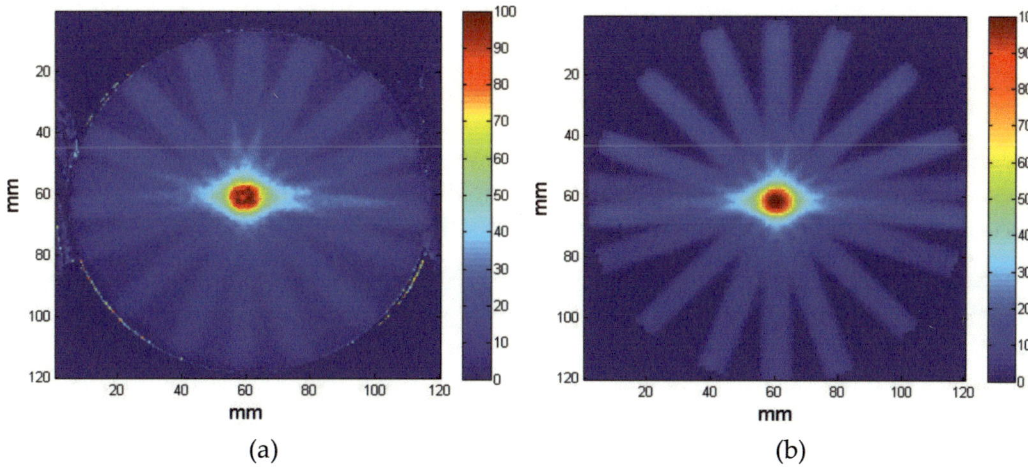

Figure 11. Dose-distributions for a conformal irradiation: (a) MAGIC-f dosimeter and (b) PENELOPE code

The comparisons between gel and PENELOPE showed maximum differences of 0.91% inside the isodose of 50% .For isodose major of 50% the difference was up to 12.3%.

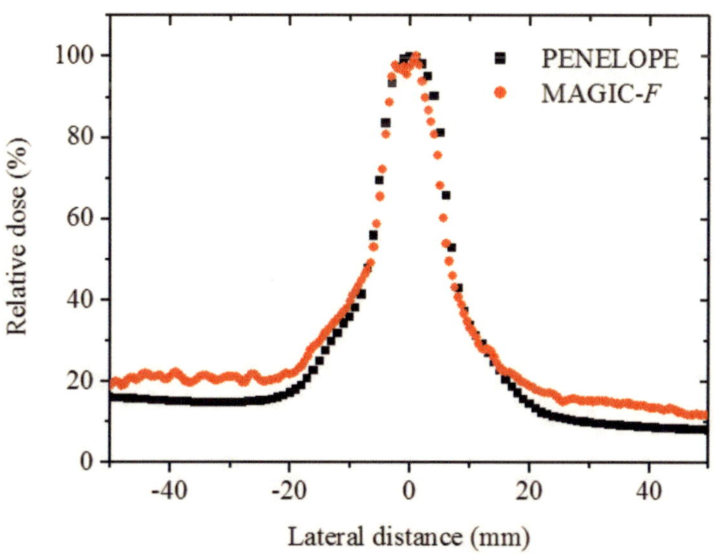

Figure 12. Beam profiles obtained with the PENELOPE code and the Magic-f gel.

From the results obtained it can be infer that the response of the MAGIC-*f* polymeric gel is suitable and can be used routinely as a dosimeter in clinical procedures using small fields, especially in radiosurgery.

4.3. Gamma index comparison for the conformal dose-distribution trough dosimetric tools

The modern radiotherapy imposes high level of accuracy to deliver absorbed-dose in a complex way, which depends on beam energy, field size, geometry of the irradiation target and others aspects. Thus, the generated dose-distributions must be checked rigorously, both for administered and delivered dose distributions.

MAGIC-f gel with high spatial resolution, effective-atomic-number close to water and an ability to measure the dose in 3D and also, PENELOPE code are available with a high concordance and estimate measurement of dose distribution in 3D.

Comparisons of dose-distributions obtained with the simulation and experimental measures can be analyzed through a beam profile, as showed in the section 4.2, which give us a limited information of that comparison. A subtraction pixel by pixel of the dose map technique can offer more information but is not related the tolerance of the position. The gamma index is a technique that relates to tolerance with the dose and position within a limited range [47,48]. Dose-distributions obtained with the MAGIC-f gel and PENELOPE simulation were compared through the technique of gamma index [49].

4.3.1. Materials and methods

The dose-distributions obtained with the MAGIC-f gel and PENELOPE code were compared by determining gamma index (γ) test. A program was developed in MalLab®, which calculated the gamma index; following the parameters of distance-to-agreement (DTA) and the tolerance of dose difference. The DTA determines the comparison tolerance of localization of one pixel from the MAGIC-f dose-distribution to the pixels within circumference, for an established radius, from the PENELOPE dose-distribution. Likewise, the dose tolerance parameter determines the comparison tolerance of dose value of one pixel from the MAGIC-f dose-distribution to the dose values to pixels within circumference, for an established radius, from the PENELOPE dose distribution.

Thus, the formulation for determining the γ values, combining the two criteria: DTA and dose tolerance, are expressed in the equation 3:

$$\Gamma = \sqrt{\frac{r^2(r_m, r)}{\Delta d_M^2} + \frac{\delta(r_m, r)}{\Delta D_M^2}} \qquad (3)$$

Since:

$r\ (r_m, r) = \left| r - r_m \right|$ is the difference between the position of a reference point of PENELOPE dose distribution, r_m, with the, r, point of the Magic-f dose-distribution, that is being compared.

$\delta(r_m,r) = D(r) - D_m(r_m)$ is difference of doses in the position r and r_m.

$\Delta d_M^2 e\ \Delta D_M^2$ are the criteria of DTA and dose difference, respectively.

Thus, to quantify the quality of the γ index for any point within range of r-r_m is chosen the minimum value of that range, which express by equation 4.

$$\gamma(rm) = \min\{\Gamma(rm,r)\}\forall\{r\} \tag{4}$$

the γ test indicate that the compared of dose distribution can be consider similar into of the criteria of the DTA and dose difference or not:

$\gamma(r_m) \leq 1$, indicate that the comparison is acceptable

$\gamma(r_m) > 1$, indicate that the comparison not is acceptable

4.3.2. Results and conclusions

As shown in the section 4.4, we obtained dose-distributions for the conformational irradiation of 9 fields of 1x1 cm², using the Magic-f gel and the PENELOPE code.

Both dose-distributions were compared using the gamma index test through the program developed using the criteria of DTA of 3 mm and dose tolerance of 3 %. Figure 13 show the γ map, with maximum values up to 2.5.

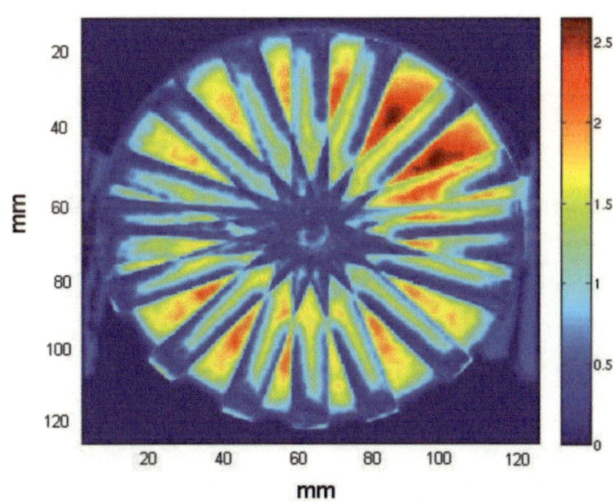

Figure 13. Gamma index value, obtained from comparing dose-distribution between Magic-f and PENELOPE.

Analyzing a central region, target volume, of 1 cm diameter from the dose-distribution obtained by the gel and PENELOPE code, 100% of the dose values pass the test under gamma index conditions. Although for regions farther than 1 cm only 57% of the dose values pass this test, which present gamma index values more than 1.

The results have shown that the response of MAGIC-f gel has a concordance with the simulated values in the central region. To improve the gel response some studies will be done to minimize the difference in region with high dose gradient and low doses to that hereafter it can be used like a dosimeter routinely in clinical procedures.

4.4. Monte Carlo simulation of MAGIC-*f* gel for radiotherapy using PENELOPE

A way to study and predict MAGIC- *f* gel results is the use of computational simulation as the Monte Carlo method. From the codes more commonly used in radiotherapy is the PENELOPE code. This code allows the "construction" of materials trough the use of the compound's chemical composition (*i.e.,* elements present and stoichiometric index, or weight fraction, of each element), mass density, mean excitation energy and energy and oscillator strength.

The aim of this application was use PENELOPE to simulate MAGIC- *f* gel dosimetric properties and to provide a valuable enhancement in the pre-constructed list of materials from PENELOPE. For validation the simulated material file (*MAGIC-f.mat*) the percentage depth doses (PDD) curves and a dose distribution were used, comparing experimental values obtained with MAGIC- *f* gel dosimeter and simulated results [50].

4.4.1. Materials and methods

The components of MAGIG- *f* gel (water, gelatin, methacrilic acid, copper sulfate, ascorbic acid and formaldehyde) and its characteristics, like atomic number, density and molar mass, were input into the PENELOPE 2008 code to build the *MAGIC-f.mat* and *water.mat* material files and for the simulation of PDDs and the conformal treatment. The simulations used 2 x 10^9 primary particles and 0.01mm^2 pixel size.

For experimental measurements were used MAGIC- *f* dosimeter. Relaxometry, weightened in T2, with a 3.0 T, NMR Philips tomography and head coil were used for the readings of the gel samples. The images were acquired with a multi spin-echo sequence with 16 echo times, TE=20ms, TR=4000ms, matrix of 256 x 256 pixels, slices of 3mm thickness. The MR images were processed and analysed with a program developed in MatLab®, that produces R2 maps.

PDDs and dose-distribution

Simulated PDDs were used to evaluate the *MAGIC-f.mat* constructed by the code. Simulation conditions were 10 x 10 cm^2 field size and 100 cm source-skin distance (SSD) for photon beams of 6 MV and 10 MV and phantoms of 30 x 30 x 20 cm^3 "filled" with both simulated materials.

Experimental PDDs were determined using MAGIC-*f* gel and ionization chamber (IC) of 0.6 cc for 6 MV and 10 MV beams. The measurements with the gel were performed using glass tubes of 16 cm length and 1 cm diameter filled with gel. The irradiations were carried out using PPMA cube filled with water and the glass tubes positioned in the center of the volume. A dose of 10 Gy in conventional irradiation conditions for both beams was used.

For the measurement of a coplanar dose-distribution, the gel was poured into a PPMA cylindrical phantom of 10 cm diameter and 15 cm height. The irradiation of the phantom was made with 5 fields of 1 x 1 cm^2, with a Varian 2100 linear accelerator gantry angles of 0° to 180°. The target was chosen in the center of the cylinder, at 100 cm SSD. The phantom was irradiated with 10 MV beam with2 Gy per field. The conformal irradiation was simulated using PENELOPE code in the same irradiation conditions. Both distributions were also compared through the dose profile.

4.4.2. Results and discussions

PDDs and dose-distribution

Figure 14 shows the PDDs through the simulations and from experimental measurements, showing them similar behavior. A maximum difference of 1.93% and 1.88% was found for the 6 MV and 10 MV beams, respectively, when simulated PDDs with *MAGIC-f.mat* and *water.mat* are compared.

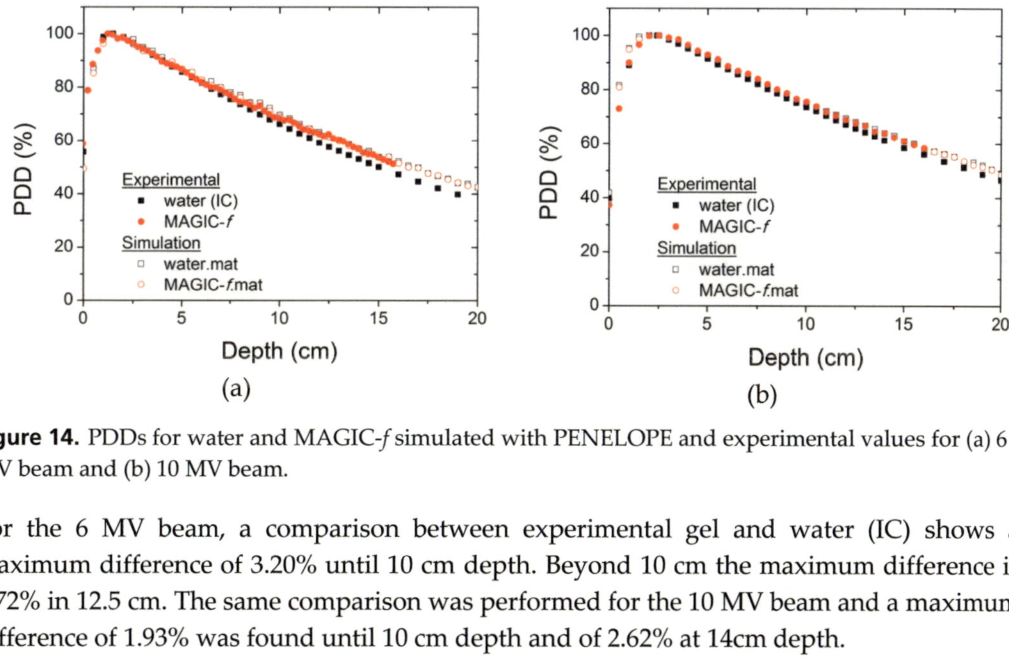

Figure 14. PDDs for water and MAGIC-*f* simulated with PENELOPE and experimental values for (a) 6 MV beam and (b) 10 MV beam.

For the 6 MV beam, a comparison between experimental gel and water (IC) shows a maximum difference of 3.20% until 10 cm depth. Beyond 10 cm the maximum difference is 5.72% in 12.5 cm. The same comparison was performed for the 10 MV beam and a maximum difference of 1.93% was found until 10 cm depth and of 2.62% at 14cm depth.

The maximum differences between experimental and simulated values are 1.46% (0.5 cm depth) and 4.3% (17cm depth) for 6 MV beam. Maximum differences of 2.14% (1 cm depth) and 3.8% (18cm depth) were found for the 10 MV beam.

The study of conformal dose distributions for 10 MV using 5 fields was performed through simulation using *MAGIC-f.mat*, shown in the figure 15.a. and experimentally using MAGIC-*f* gel dosimeter as shown in the figure 15.b.

Figure 16 shows the dose profiles obtained from both distributions.

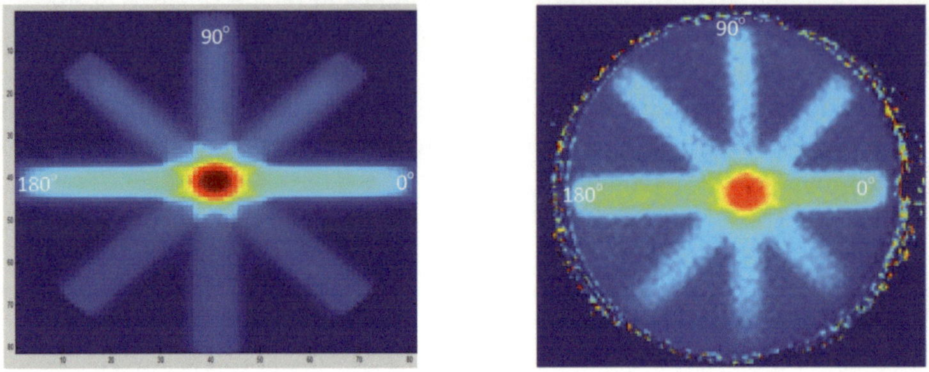

(a) (b)

Figure 15. Dose distribution for 10 MV obtained with: (a) PENELOPE; (b) MAGIC-f dosimeter.

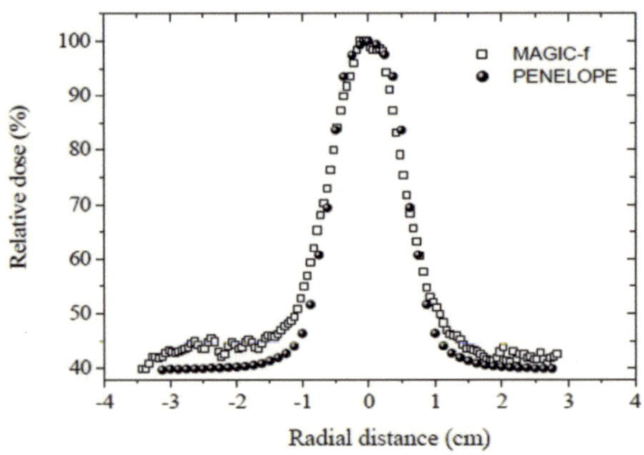

Figure 16. Dose profiles of simulated and experimental distributions.

A maximum difference of 0.5% in a radial distance of 0.55 cm inside the field (approximately half the irradiation field) was found and beyond that distance the differences between the dose distributions are less than 7%.

The simulations using PENELOPE code showed similar behaviors with the experimental values. Although discrepancies were found, specifically in the dose-distribution in regions of low doses, this study shows that the code can be used to simulate the components of the MAGIC-f gel for these energies. More over, the *MAGIC-f.mat* file can be useful for studies of response and comparison of the gel.

5. Conclusions

The validation of the MAGIC-f gel and PENELOPE through the dosimetric parameters, percentage depth dose and beam profile showed that that both dosimetric tools present similar behavior.

The use of two dosimeters for all radiotherapy modalities present in this work showed great concordance, although discrepancies were found in experimental measurements, using the MAGIC-f gel, and the simulations obtained with the PENELOPE code.

From the results obtained it can be infer that the response of the MAGIC-*f* polymeric gel and simulation by PENELOPE-Monte Carlo code are suitable and can be used routinely as a dosimeter, in case of the gel, and in the way to study or predict the response and shape of dose-distribution by the simulation.

Author details

Mirko Salomón Alva-Sánchez and Thatiane Alves Pianoschi
Department of Physics, University of São Paulo, Av. Bandeirantes 3900 CEP:14010-901- Ribeirão Preto – SP, Brazil

Acknowledgement

The presented works were supported by: CAPES (*Coordenação de Aperfeiçoamento de Pessoal de Nível Superior*) and the physics department of the University of Sao Paulo-Campus Ribeirao Preto-Brazil.

We are thankfully to the personnel from the Cancer Hospital of Barretos, especially to Marcelo Carvalho Santanna. Hospital of Clinics from Ribeirão Preto, Hospital Sírio Libanês-Sao Paulo-Brazil, especially to Cecilia Hadad.

The authors are thankfully to the professor Ph. D. Patricia Nicolucci to orientation and suggestions for all the works, which are part of our doctorate work.

We are thankfully for technical support to José Luiz Aziani and Carlos Renato da Silva is also appreciated.

6. References

[1] International Commission radiation and units, ICRU, (1999) Prescribing Recording and Reporting Photon Beam Therapy (Supplement to ICRU Report 50), Report n° 62, ICRU, Bethesda, 53 p.

[2] American Association of Physics in Medicine, AAPM, (1994) Code of practice for radiotherapy accelerator: Report of AAPM radiotherapy task Grop no 45. Med. Phys. 21:1093-1121.

[3] International atomic energy agency, IAEA, (2000). ''Absorbed dose determination in external beam radiotherapy: An international code of practice for dosimetry based on standards of absorbed dose to water '', thechnical Report Series. IAEA TRS 398, IAEA, Vienna, Austira.

[4] De Deene Y (2002) Gel dosimetry for the dose verification of intensity-modulated radiotherapy treatments Z. Med. phys. 12: 77–88.

[5] De Deene Y, Hurley C, Venning A, Vergote K, Mather M, Healy B, Baldock C (2002) A basic study of some normoxic polymer gel dosimeters. Phys. med. biol. 47: 3441–3463.

[6] McJury M, Oldham M, Cosgrove V P, Murphy P S, Doran S, Leach M, Webb S (2000) Radiation dosimetry using polymer gels: methods and applications. Br. J. Radiol. 73: 919–929.

[7] Haraldsson P, Karlsson A, Weislander E, Gustavsson H, Bäck S (2006) Dose response evaluation of a low-density normoxic polymer gel dosimetry uising MRI. Phys. med. biol. 51: 919-928.

[8] Pavoni J, Pike T, Snow J, DeWerd L, Baffa O (2010) Dosimetria tridimensional usando gel MAGIC com formaldeído. Rev. bra. de física médica. 4: 15-18.

[9] Zhu X, Fakhri G, Reese T, Crowley E (2010) Improved MAGIC gel for higher sensitivity and elemental tissue equivalent 3D dosimetry. Med. phys. 37: 183-189.

[10] Pianoschi T, Alva M, Santanna M, Baffa O, Nicolucci P (2010) Magic-f gel dosimeter for clinical electron beam. J. phys. conf. ser. 250: 012037.

[11] Schwarcke M, Marques T, Garrido C, Nicolucci P, Baffa O (2010) MAGIC-f gel in Nuclear Medicine Dosimetry: study in an external beam of Iodine- 131. J. phys. conf. ser. 250: 012082.

[12] De Vlamynck K, Palmans H, Verhaegen F (1999) Dose measurements compared with Monte Carlo simulations of narrow 6 MV multifead collimator shaped photon beams. Med. phys. 26: 1874-1882 .

[13] Yamamotro T, Mizowaki T, Miyabe (2007) An integrated Monte Carlo dosimetric verification system for radiotherapy treatment planning. Phys. med. bio. 52: 1991-2008.

[14] Al-Ghorabie F, Natto S, Al-Lyhiani S (2001) A comparison between EGS4 and MNCP computer modeling of an in vivo X-ray fluorescence system. Comp. bio. med. 31: 73-83.

[15] Chiavassa S, Lemosquet, Aubineau-Lani L, de Carlan L, Clairand I, Ferrer L, Bardi M, Franck D, Zankl M (2005) Dosimetric comparison of Monte Carlo codes (EGS4, MCNP, MCNPX) considering external and internal exposures of the zubal phantom to electron and photon source. Radiat. protec. Dosim. 116: 631–635.

[16] Neto A, Haddad C, Pelosi E, Zevallos-Chávez J, Yoriyaz H, élio Yoriyaz, Siqueira P (2005) Monte Carlo Simulation as an Auxiliary Tool for Electron Beam Quality Specification for Intra-Operative Radiotherapy. Braz. J. phys. 35: 801.

[17] Sempau J, Badal A, Brualla L (2011) A PENELOPE-based system for the automated Monte Carlo simulation of clinacs and voxelized geometries—application to far-from-axis fields. Med. phys. 38: 5887-5895.

[18] Koivunoroa H, Siiskonen T, Kotiluoto P, Auterinen I, Hippelainen E, Savolainen S (2012) Accuracy of the electron transport in MCNP5 and its suitability for ionization chamber response simulations: A comparison with the EGSNRC and PENELOPE codes. Med. phys. 39:1335-1344.

[19] Ramirez J, Chen F, Nicolucci P, Baffa O (2011) Dosimetry of small radiation field in inhomogeneous medium using alanine/EPR minidosimeters and PENELOPE Monte Carlo simulation. Radiat. Measur.46:941-944.

[20] Gonzales D, Requena S, Williams S (2012) Au La x-rays induced by photons from 241Am: Comparison of experimental results and the predictions of PENELOPE. Appl. Radiat. Isot. 70: 301–304.

[21] Alexander P, Charlesby A, Ross M (1954) The degradation of solid polymethylmethacrylate by ionization radiation. Proc. r. soc. A: 223-392.

[22] Boni A (1961) Polyacrylamide gamma dosimeter. Radiat. Res. 14: 374-380.

[23] Audet C, Schreiner L (1991) Radiation dosimetry by NMR relaxation time measurement of irradiated polymer solutions. Proc. Soc. magn. reson med. 10th Annual Scientific Meeting 705.

[24] Maryanski M, Gore J, Kennan R, Schulz R (1993) NMR relaxation enhancement in gels polymerized and cross-linked by ionizing radiation: a new approach to 3D dosimeter by MRI. Mag. reson. imaging. 11: 253-258.

[25] Maryanski M, Schultz R, Ibbott G, Gatenby J, Xie J, Horton D, Gore J (1994) Magnetic resonance imaging of radiation dose distributions using a polymer-gel dosimeter. Phys. med. biol. 39: 1437-1455.

[26] Baldock C, Burford R, Billinghan N, Wagner G, Patval S, Badawi R, Keevil S (1998) Experimental procedure for the manufacture of polycrylamide gel (PAG) for magnetic resonance imaging (MRI) radiation dosimetry. Phys. med. biol. 43: 695-702.

[27] Fong P, Keil D, Does M, Gore J (2001) Polymer gels for magnetic resonance imaging of radiation dose distributions at normal room atmosphere. Phys. med. biol. 46: 3105-3113.

[28] Baldock C, Burford R, Billinghan N, Wagner G, Patval S, Badawi R, Keevil S (1998) Experimental procedure for the manufacture of polycrylamide gel (PAG) for magnetic resonance imaging (MRI) radiation dosimetry. Phys. med. biol. 43: 695-702.

[29] De Deene Y, Hurley C, Venning A, Vergote K, Mather M, Healy B, Baldock C (2002) A basic study of some normoxic polymer gel dosimeter. Phys. med. biol. 47: 3441-3463.

[30] Gustavsson H, Karlsson A, Back S, Olsson L, Haraldsson P, Engstrom P, Nystrom H (2003) MAGIC-type gel for three-dimensional dosimetry: Intensity-modulated radiation therapy verification. Med. phys. 30:1264-1271.

[31] Fernandes J, Pastorello B, de Araujo D, Baffa O (2008) Formaldehyde increases magic gel dosimeter melting point and sensitivity. Phys. med. biol. 53: N1-N6.

[32] INTERNATIONAL COMIMISSION ON RADIATION UNITS AND MEASUREMENTS, ICRU, (1989)Tissue Substitutes in Radiation Dosimetry and Measurement. ICRU Report 44, Washington.

[33] Salvat F, Fernandez-Varea J, Acosta E, Sempau J (2005) PENELOPE – A Code System for Monte Carlo Simulation of Electron and Photon Transport, Nuclear Energy Agency OECD/NEA, Issy-les-Moulineaux, France. Available: http://www.nea.fr. Accessed 2011 Nov 02.

[34] Yan X, Poon E, Reniers B, Voung T, Verhaegen F (2008) Comparison of dose calculation algorithms for colorectal cancer brachytherapy treatment with a shielde applicator. Med. phys. 35: 4824-4830.

[35] Takam R, Bezak E, Yeoh E (2009) Risk of secondary primary cancer following prostate cancer radiotherapy: DVH analysis using the competitive risk model. Phys. med. biol. 54: 611-625.

[36] Goggen T (1988) Physical Aspects of Brachytherapy. Medical Physics Handbooks editors, England, pp 198.

[37] Nath R, Anderson L, Meli J. Olch A, Stitt J, Williamsom J (1997) Code of practice for brachytherapy physics: Report of the AAPM Radiation Therapy Committee Task Group No. 56. Med. phys. 24:1557-1598.

[38] International Atomic Energy Agency, IAEA, (2002) Calibration of photon and beta ray sources used in brachytherapy. IAEA-TECDOC 1274.

[39] Sarfehnia A, Stewart K, Seuntjens J (2007) An absorbed dose to water standard for HDR ^{192}Ir brachytherapy sources based on water calorimetry: numerical and experimental proof-of-principle. Med. phys. 34: 4957-4961.

[40] Alva M, Marques T, Schwarcke M, Gonçalvez L, Baffa O, Nicolucci P (2009) Water-equivalent calibration of ^{192}Ir HDR Brachytherapy source using MAGIC-f gel Polymer. In: World Congress on Medical Physics and Biomedical Engineerin. IFMBE Proceedings. Munich : Springer Berlin Heidelberg. 25: 248-251.

[41] AGÊNCIA INTENACIONAL DE ENERGIA ATOMICA, IAEA, (2005) "Determinación de la dosis absorbida en Radioterapia con haces externos: un código de práctica internacional par la dosimetria basada en patrones de dosis absorbida en agua". Viena: Agencia Internacional de Energia Atômica, IAEA-informe técnico 398.

[42] Babic S, Battista J, Jordan K (2009) Three-dimensional dosimetry of small megavoltage radiation fields using radiochromic gels and optical CT scanning. Phys. med. biol. 54: 2463-2481.

[43] Wong C, Ackerly T, He C, Patterson W, Powell C, Qiao G, Solomon D, Meder R, Geso M (2009) Small field size dose-profile measurements using gel dosimeters, gafchromic films and micro-thermoluminescent dosimeters. Radiat. mea. 44: 249-256.

[44] Calcina C, Oliveira L, Almeida C, Almeida A (2007) Dosimetric parameters for small field sizes using Fricke xylenol gel, thermoluminescent and film dosimeters, and an ionization chamber. Phys. med. biol. 52: 1431-1439.

[45] Das J, Ding G, Ahnesjö A (2008) Small fields: Nonequilibrium radiation dosimetry. Med. phys. 35: 206-215.

[46] Alva M, Pianoschi T, Takeda F, Alves T, Haddad C, Nicolucci P. (2010) Dose Distribution of Small Fields through MAGIC-F Gel Dosimetry and PENELOPE-Monte Carlo Simulation. Med. phys. 37: 3122.

[47] Low D, Harms W, Mutic S, Purdy J (1998) A technique for the quantitative evaluation of dose distributions. Med. phys. 25:656-661.

[48] Low D, Dempsey J (2003) "Evaluation of the gamma dose distribution comparison method. Med. phys. 9:2455–2464.

[49] Alva M, Pianoschi, Amaral L, Oliveira H, Nicolucci P(2011) Gamma Index comparison for the conformational dose distribution obtained through three dosimetric tools. Braz. J. Med. phys. 5: 66.

[50] Alva M, Pianoschi T, Marques T, Santanna M, Baffa O, Nicolucci P (2010) Monte Carlo Simulation of MAGIC- gel for Radiotherapy using PENELOPE. J. phys. conf. ser. 250: 012067.

Chapter 5

A Respiratory Motion Prediction Based on Time-Variant Seasonal Autoregressive Model for Real-Time Image-Guided Radiotherapy

Kei Ichiji, Noriyasu Homma, Masao Sakai, Makoto Abe, Norihiro Sugita and Makoto Yoshizawa

Additional information is available at the end of the chapter

http://dx.doi.org/10.5772/56554

1. Introduction

In radiation therapy, to deliver continuously a sufficient radiation dose to target volume yields a better therapeutic effect. While, avoiding an exposure to healthy tissues surrounding the target volume is also an important requirement for suppressing the adverse effect. Image-guided radiation therapy (IGRT) has potential to achieve the two requirements and as it's application, stereotactic body radiation therapy (SBRT) has been used in clinic. In SBRT, the irradiated field is positioned with millimeter accuracy by proper daily setup. The accurate irradiation can allow the increase of radiation dose by ignoring the irradiation to the healthy tissues. Indeed, it has been reported that the treatment result of SBRT is comparable to the outcome from surgery [1].

As mentioned above, the higher accuracy as irradiation allows the higher planning dose to the target. However, intra-fractional organ motion, such as respiratory motion of lung, often makes misalignment of the isocenter and the target volume during treatment fraction. For example, respiratory motion moves lung tumor over 10 mm per second [2, 3]. In this case, the irradiation error caused by intra-fractional motion is not negligible in terms of adverse effect. Therefore, the respiratory motion management is a very important task in the field of radiation treatment [4].

To take into account the intra-fraction motion of lung tumor, some irradiation techniques have been investigated. The technique most widely used is the use of internal target volume (ITV) [5]. ITV includes internal margin determined by the intra-fractional organ motion. Consequently ITV covers the target volume CTV without misalignment as shown in Figure 1). However, radiation dose must be lower than the case of the irradiation without internal margin, because the irradiated area also covers normal tissues surrounding the target volume. On the other hand, a gating technique can give a high radiation dose to the lung tumor [6, 7].

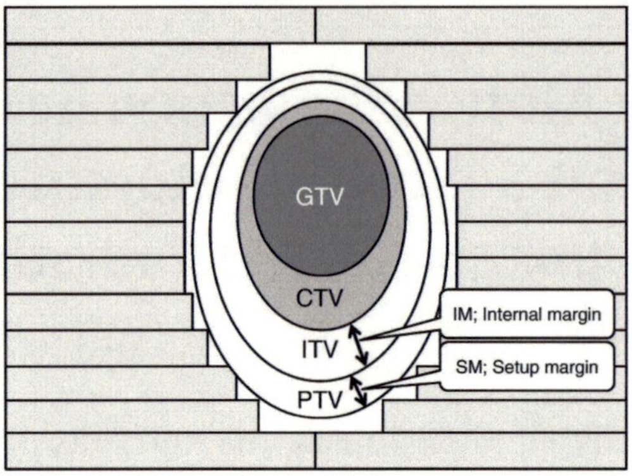

Figure 1. A relationship between target volumes. GTV; gross tumor volume. CTV; clinical tumor volume. ITV; internal target volume. PTV; planning target volume.

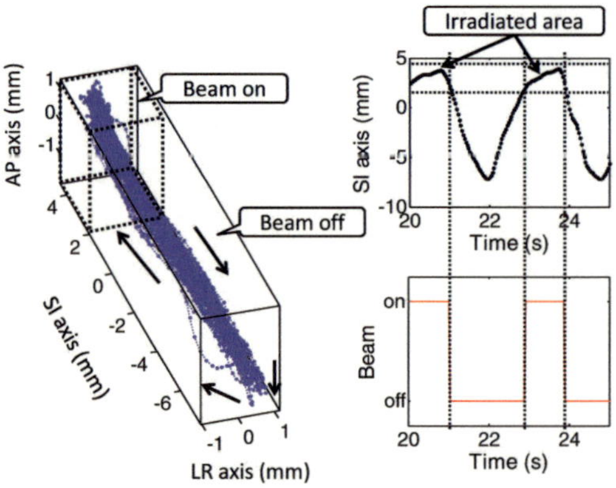

Figure 2. A schematic diagram of gated radiotherapy. The dots denotes a three-dimensional tumor trajectory. Irradiation is interrupted when the tumor is out of the region drawn as dotted line. AP, SI, and LR denotes anterior-posterior, superior-inferior, and left-right axes, respectively.

This is on-and-off irradiation to static region as shown in Figure 2. The radiation dose is delivered when the target volume is within the area planned preliminarily, and the irradiation interrupts if the target volume is out of the planned area. The gating technique can suppress the exposure to healthy tissues and allows high-dose-rate irradiation which is sufficient to yield a better therapeutic effect. However, instead of high-dose exposure, the gating takes longer treatment time to yields same therapeutic effect to SBRT due to the interruption.

An ideal irradiation to the moving target is to continuously irradiate a sufficient dose to the tumor which can be achieved by controlling the radiation beam to chase the moving target [8]. Such real-time tumor following (or chasing) irradiation yields an ideal therapeutic effect and can shorten the treatment duration. A schematic diagram of the tumor following irradiation is shown in Figure 3.

The tumor following irradiation requires the following two key techniques at least.

(i) Real-time measurement of tumor position and shape.

(ii) Real-time repositioning and reshaping the treatment beam

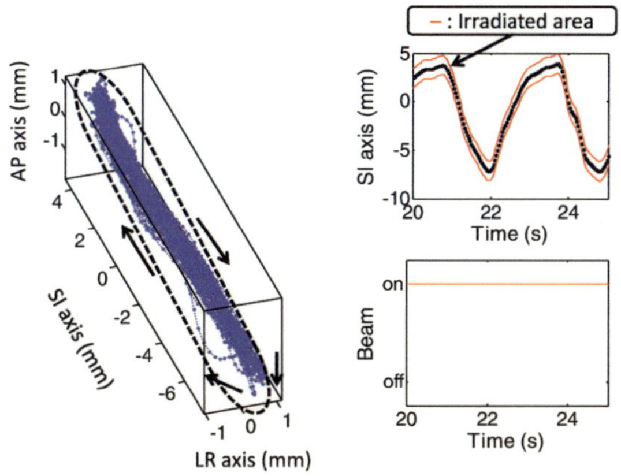

Figure 3. A schematic diagram of tumor following radiotherapy. The dots denotes a three-dimensional tumor trajectory. Moving tumor is continuously irradiated as the region drawn as dotted line. AP, SI, and LR denotes anterior-posterior, superior-inferior, and left-right axes, respectively.

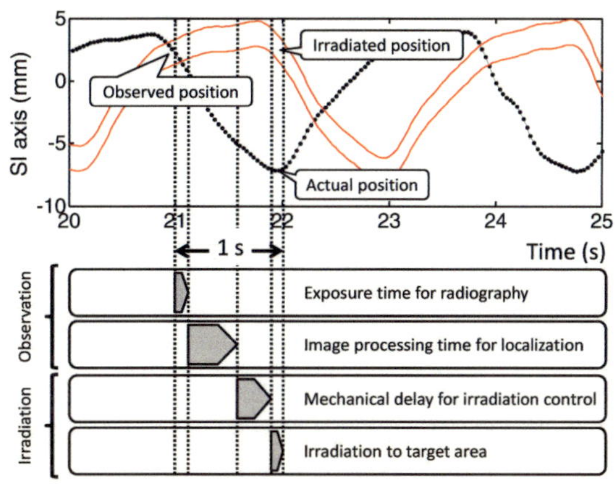

Figure 4. A schematic diagram of the tumor following irradiation with system latency. In this example, the total delay is 1 s and the delay distances the target tumor from the irradiated area over 10 mm.

The first technique for tumor localization has been achieved by using an X-ray fluoroscopic imaging system such as real-time tumor tracking system [7]. The second technique of beam-positioning systems can be realized by using dynamic multi-leaf collimator, robotic couch, robotic manipulator, and so on. However, the tumor following irradiation has not been developed clinically yet. This is because even we can obtain the precise lung tumor position in real-time, current radiotherapy machine has mechanical and computational time delays of up to about 1 s for controlling the irradiation field and image processing [9]. The latency can definitely affect badly on the irradiation accuracy [10], and must be compensated. Figure 4 shows a schematic diagram of the tumor following irradiation without delay compensation.

Prediction of future lung tumor motion is a typical solution for compensating the system latency and thus is a hot research topic for the tumor following irradiation. Then, the third technique is needed to develop the tumor following radiation therapy.

(iii) Real-time motion prediction

Several prediction methods have thus been proposed for the respiratory motion. These include linear regression [11, 12], extrapolation [11], artificial neural network [11, 13], Kalman filter [11], nonlinear regression based on the Takens theorem [12, 14], probabilistic modeling [15, 16] and so on. More details can be found in a literature such as a survey study of the prediction of lung tumor motion [17]. A variety of the prediction approaches indicates that there is currently no best prediction method in clinical use because of the insufficient accuracy.

In general, the prediction is realized as an application of time series analysis. General prediction methods, such as autoregressive moving average (ARMA) model [18], require a stationarity of the target time series. However, the respiratory motion essentially has complex and time-varying characteristics. For example, the repetition of inhalation and exhalation naturally involves periodicity, but the periods are time-varying. Nevertheless the periodic component can still help to predict the motion because the past observed motion will arise again at periodic intervals. It means that we can predict the motion accurately if the period of target motion is obtained. Thus, the use of the periodicity can be a good approach to predict the respiratory motion.

The methods that focus on periodic components in respiratory motion have been developed as seasonal autoregressive (SAR) model-based method [19] and periodic ARMA model-method [20]. However, the periodicity in the respiratory motion fluctuates with time, and the adaptation of these methods to time-variant periodic variation seems insufficient yet. For example, SAR model-based method converts the target motion with time-variant period to a new motion with constant period to use the SAR model properly. But if the conversion from time-variant to constant period is incomplete, the prediction accuracy can decrease. On the other hand, the periodic ARMA estimates the period of the target motion by using long historical samples of 60 s. However, the use of long historical samples causes a hysteresis and cannot trace the change of the periodicity. Therefore, to adapt to the time-varying periodicity still remains as a challenge to accurately predict the respiratory motion.

In this chapter, to predict the respiratory motion by use of its periodicity, a time-variant seasonal autoregressive (TVSAR) model-based method is proposed [21]. The proposed method can model the time-variant periodicity more precisely, and we adopt only a small number of samples to suppress the hysteresis. In the next section, we briefly describe the characteristics of the lung tumor motion. Then, in section 3, TVSAR model is explained. The model is based on SAR, but is newly developed to adapt to the fluctuated periodicity by using unequally-spaced intervals instead of multiples of the constant period. To show the prediction performance of TVSAR model-based method, experimental result by using some clinical data sets of lung tumor motion is described in section 4. Section 5 provides concluding remarks.

2. Respiratory motion of lung tumor

Figure 5 shows an example of time series of respiratory motion of lung tumor. The time series of tumor motion was observed as a location of the golden fiducial marker implanted into around the tumor, by using a kV X-ray fluoroscopic imaging system known as RTRT system, with sampling frequency 30 Hz, at Hokkaido University Hospital, Sapporo, Japan [7].

In general, a lung tumor motion involves a periodical component because breathing is composed of repetition of inspiration and expiration. Figure 6 provides power spectrum density (PSD) of the tumor motion shown in the top of Figure 5. The dominant frequency

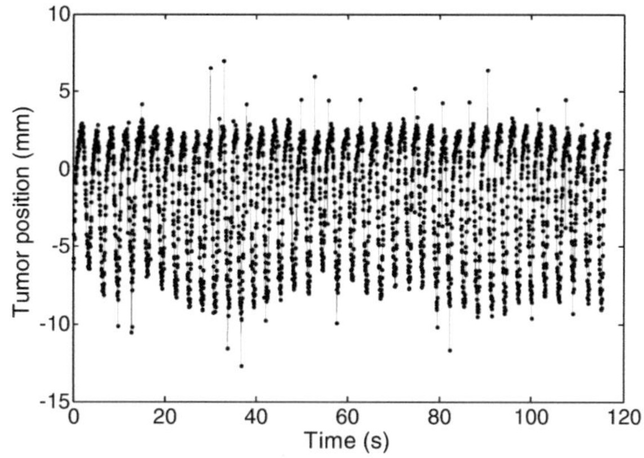

Figure 5. An example of time series of lung tumor motion.

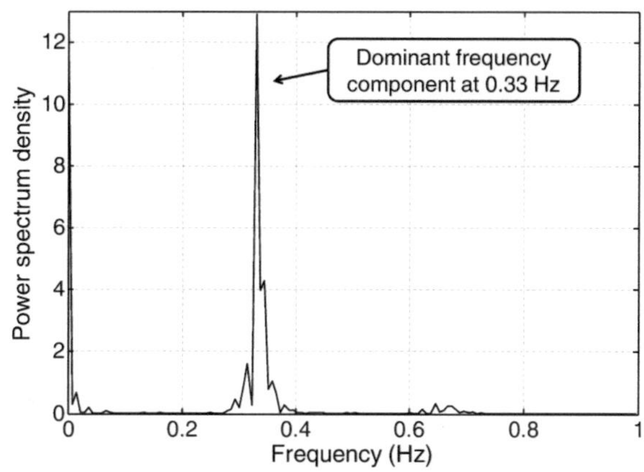

Figure 6. Power spectrum density of the respiratory induced tumor motion shown in Figure 5. The dominant frequency component is found at 0.33 Hz.

corresponding to the respiratory cycle can be found at 0.33 Hz approximately in this example. This means that the tumor motion has a periodical component induced by the 3 s respiratory cycle in average.

As is clear from the analysis, one of dominant components in respiratory motion is periodic variation due to repetition of inhalation and exhalation. However, the period of the motion is not constant, but a time varying function. Such quasi-periodical nature of the motion is simply found as fluctuation of peak-to-peak intervals. For example, time interval between a peak and next peak differs each other even if the amplitude variation is sufficiently small as shown in Figure 7. This suggests that the dominant frequency is time-varying.

Recalling that the breathing period is time-variant, short-time Fourier transform (STFT), instead of the normal long-time Fourier transform, was performed on the motion example. The time-variation of the frequency spectrum is shown in Figure 8. The dominant frequency at each time is depicted as black dashed line. The line clearly indicates that the respiratory cycle fluctuates with time. The range of the fluctuation was from 0.298 to 0.360 Hz in this example.

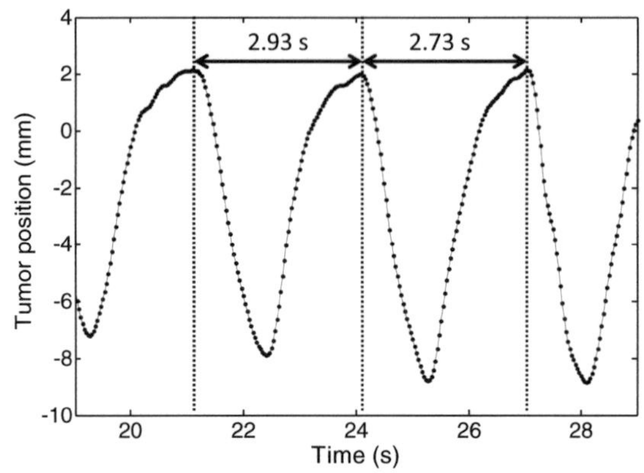

Figure 7. An example of fluctuated periodicity of respiratory motion. The peak to peak intervals indicated as double-headed arrows are not same to each other.

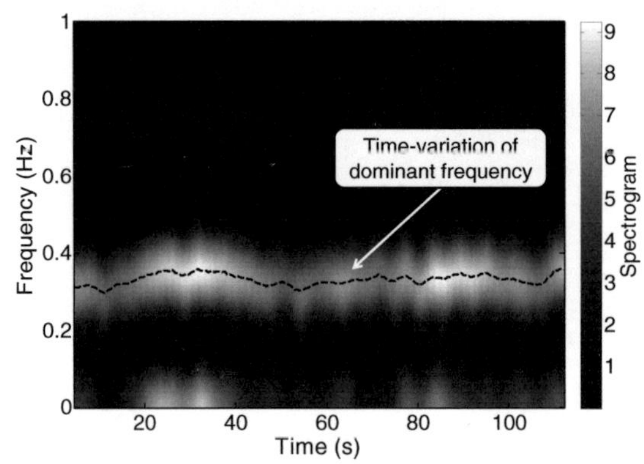

Figure 8. Time-variation of power spectrum density. The dominant frequency fluctuates with time. The range of fluctuation is from 0.298 to 0.360 Hz in this example.

3. Prediction methods

The periodicity found in the respiratory motion is a typical one of nonlinear and complex nature but also it can be useful to predict itself. That is, the past observed patterns in the motion will repeatedly arise at constant period. In this section, seasonal autoregressive (SAR) model is explained as a method with use of periodicity. Also, a limitation of general SAR model for lung tumor motion prediction is shown. Then, time-variant SAR model is introduced as a method designed for tumor motion prediction.

3.1. Seasonal autoregressive (SAR) model

Seasonal autoregressive integrated moving-average (SARIMA) model [22] is a general expression of the time series which changes periodically (i.e., a periodical function of time such that $x(t) \sim x(t - \rho \cdot s)$).

The SARIMA model of the time series $\{x(0), x(1), \ldots, x(T)\}$ with period s sample can be expressed as follows.

$$\phi(B)\Phi(B^s)(1 - B)^d(1 - B^s)^D x(t) = \theta(B)\Theta(B^s)\epsilon(t) \tag{1}$$

where d and D are respectively the order of local and seasonal integrated components, $\epsilon(t)$ is the Gaussian noise of which mean and variance are 0 and σ^2, respectively, and B is a delay operator defined by

$$B^k x(t) = x(t-k),\ k = 1, 2, \ldots \tag{2}$$

Then, each components of the SARIMA model are given as follows.

$$\text{Autoregressive: } \phi(z) = 1 - \phi_1 z - \cdots - \phi_p z^p \tag{3}$$

$$\text{Moving-average: } \theta(z) = 1 + \theta_1 z + \cdots + \theta_q z^q \tag{4}$$

$$\text{Seasonal AR: } \Phi(z) = 1 - \Phi_1 z - \cdots - \Phi_P z^P \tag{5}$$

$$\text{Seasonal MA: } \Theta(z) = 1 - \Theta_1 z + \cdots + \Theta_Q z^Q \tag{6}$$

where p, q, P and Q are the orders of four components in (3)-(6) respectively.

The SARIMA model can express various periodical time series by designing the model parameters. In the followings, let us consider only the seasonal autoregressive (SAR) component to avoid the over fitting problem and to simplify the explanation of the prediction method. That is, let $p = q = Q = d = D = 0$ in (1), then for this special case, we can obtain SAR model for the time series $y(t)$ as follows.

$$y(t) = \epsilon(t) + \sum_{\rho=1}^{P} \Phi_\rho \cdot y(t - \rho \times s) \tag{7}$$

where P is the order of the SAR components, $\Phi_\rho, \rho = 1, 2, \ldots, P$ are the SAR coefficients, and s are the constant period of the target time series.

Then, to substitute $t + h$ for t, the prediction equation by using (7) can be given by

$$\hat{y}(t+h|t) - \sum_{\rho=1}^{P} \Phi_\rho \cdot y(t + h - \rho \times s) \tag{8}$$

where $\hat{y}(t+h|t)$ is the predicted value of future time $t+h$ samples with h samples ahead of current time t, and the term $-\rho \times s + h$ must be not longer than 0 for composing the prediction by using only past observations.

The equations (7) and (8) show that an essential core of the general SAR depends on an assumption that each values of the same phase correlate each other. These equations can provide a simple description with a single model for the periodical variation with period s. For example, if target time series are given by the following deterministic and fully periodical function:

$$x(t) = A_0 + \sum_{n=1}^{N} A_n \cdot \cos\left(2\pi \frac{n}{s} t - \varphi_n\right) \tag{9}$$

where $A_n, n = 0, 1, 2, \ldots$ and $\varphi_n, n = 1, 2, 3, \ldots$ are amplitudes and initial phases for each n-th harmonic component, respectively. Then, let $\Phi_\rho = 1/P$ in Eq. (8), the SAR model can well

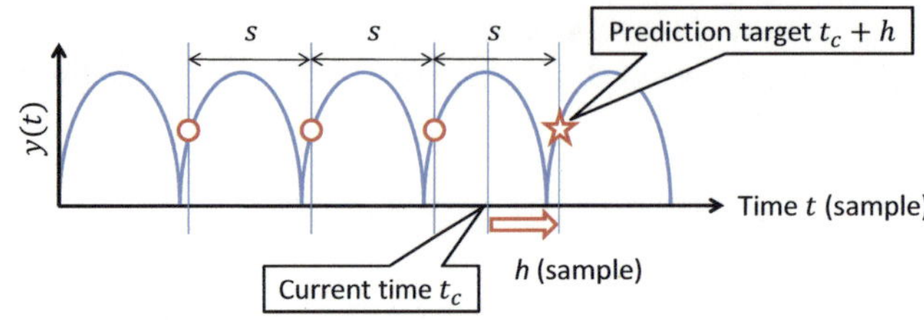

Figure 9. Schematic diagram of SAR model-based prediction for time series with constant period.

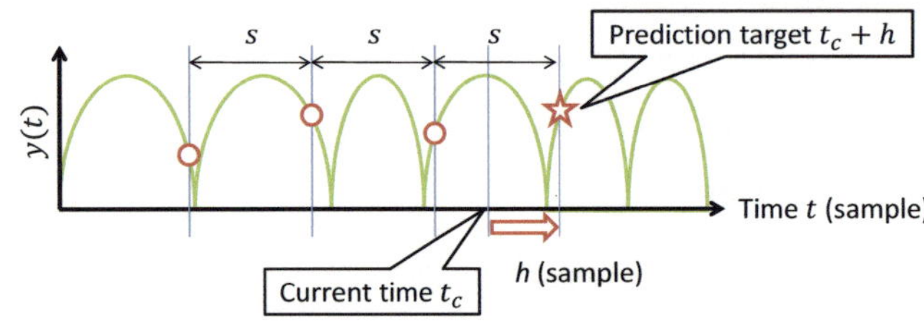

Figure 10. Schematic diagram of SAR model-based prediction for time series with fluctuated period.

predict the future value of the periodical function $x(t+h)$ at h-sample ahead future as follows.

$$\begin{aligned}
&\hat{x}(t+h|t) \\
&= \frac{1}{P}\sum_{\rho=1}^{P}\left(A_0 + \sum_{n=1}^{N} A_n \cdot \cos\left(2\pi\frac{n}{s}(t+h-\rho\cdot s) - \varphi_n\right)\right) \\
&= \frac{1}{P}\sum_{\rho=1}^{P}\left(A_0 + \sum_{n=1}^{N} A_n \cdot \cos\left(2\pi\frac{n}{s}(t+h) - 2\pi\rho - \varphi_n\right)\right) \\
&= \frac{1}{P}\sum_{\rho=1}^{P}\left(A_0 + \sum_{n=1}^{N} A_n \cdot \cos\left(2\pi\frac{n}{s}(t+h) - \varphi_n\right)\right).
\end{aligned} \tag{10}$$

Figure 9 shows a schematic diagram of SAR model-based prediction for simple periodic time series with period s. In this example, SAR model can predict the target value (drawn as star) by referring the past observed values (drawn as circles) which are similar to the target value by using constant period s.

As shown by the prediction example, the SAR model-based equation (8) is useful and powerful to predict unknown future value of periodic time series. Then, remember that the dominant component in respiratory tumor motion is periodical component. These facts suggest that the SAR model is suitable to predict the tumor motion. However, remember that the periodicity in our target time series, lung tumor motion, fluctuates. The use of constant period s in equation (8) may be improper for the tumor motion. That is, if the periodical nature in the target time series changes with time, the constant period will provide a large prediction error.

A simple solution for adapting the SAR model to the fluctuated periodicity is to substitute a time-variant period $s(t) for$the constant period s given by

$$\hat{y}(t) = \epsilon(t) + \sum_{\rho=1}^{P} \Phi_\rho \cdot y(t - \rho \cdot s(t)). \tag{11}$$

Then the new prediction equation is written as

$$\hat{y}(t+h|t) = \sum_{\rho=1}^{P} \Phi_\rho \cdot y(t+h-\rho \cdot s(t+h)). \tag{12}$$

The equation above is about the same to the SAR model-based method [19] and it seems that it can adapt to the fluctuated periodicity, but this simple extension of SAR model has a limitation.

For explanation of the limitation, let us consider time series with a time-varying period as a function of time t. For this situation, the instantaneous phase $\vartheta(t)$ of the time series is given as a time-integration of the angular velocity $\omega(t)$, such as $\vartheta(t) = \int_0^t \omega(t)dt$. If the angular velocity is given as a linear function of time t,

$$\omega(t) = \omega_a t + \omega_0, \tag{13}$$

where ω_a and ω_0 are constants. And the instantaneous phase is given a nonlinear (the second-order in this case) function as

$$\vartheta(t) = \frac{1}{2}\omega_a t^2 + \omega_0 t. \tag{14}$$

Then let us assume that $s(t)$ is known or estimated correctly. That is, the term $s(t)$ can refer the instantaneous phase same to the target time t, i.e., $\vartheta(t) = \vartheta(t - s(t))$ However, the terms $\rho \cdot s(t), \rho = 2, 3, \ldots, P$ in (11) cannot refer the same instantaneous phase, i.e., $\vartheta(t) \neq \vartheta(t - \rho \cdot s(t)) + 2\pi\rho$, even if the $s(t)$ is given as a true value.

Thus, the simple SAR model-based solution by using the time-varying period $s(t)$ cannot express the fluctuated periodic nature suitably, even if the change of the periodicity is sufficiently simple such as a linear function of time. This is one of the limitations of SAR model in the prediction of fluctuated periodical time series such as respiratory tumor motion.

3.2. Time-variant SAR (TVSAR) model

To overcome the limitation of the general SAR model, a time-variant SAR (TVSAR) model for prediction of the lung tumor motion has been proposed [21]. The TVSAR equation corresponding to (7) can be expressed as

$$y(t) = \epsilon(t) + \sum_{\rho=1}^{P} \Phi_\rho \cdot y(t - r_\rho(t)). \tag{15}$$

where $r_\rho(t)$ are reference intervals of the ρth order at time t.

The reference intervals $r_\rho(t)$ can be defined as a time interval between the current time and the corresponding past time which has the same phase to the current one. In other words, if an instantaneous phase $\vartheta(t)$ of the time series is given, the reference intervals can be defined as follows.

$$r_\rho(t) = \arg\min_{k>0} |\vartheta(t) - 2\rho\pi - \vartheta(t-k)| \tag{16}$$

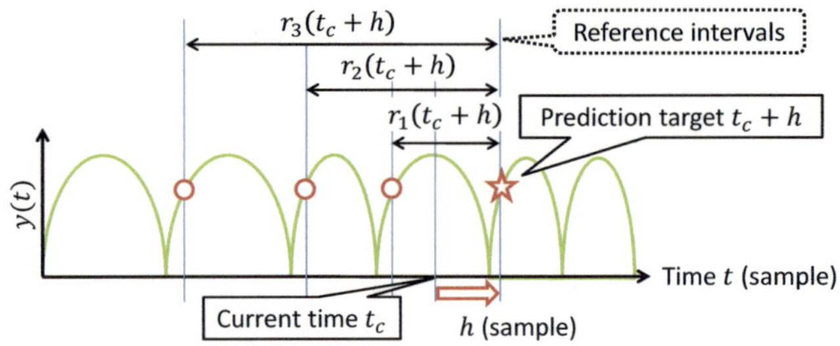

Figure 11. Schematic diagram of time-variant SAR (TVSASR) model-based prediction for time series with fluctuated period.

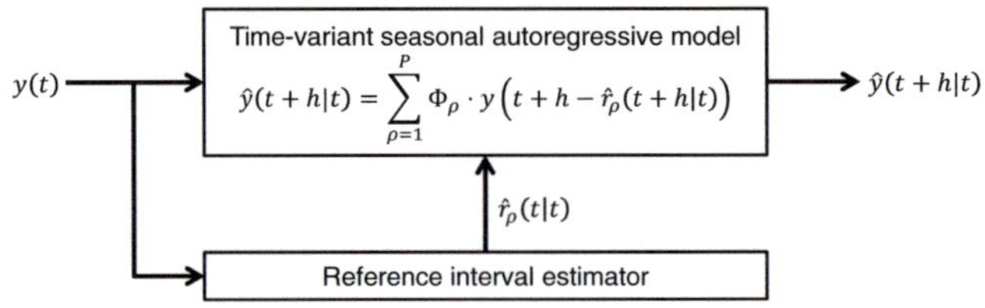

Figure 12. Schematic diagram of TVSAR model-based prediction system.

Also, if the period is a constant, the reference interval can be expressed by

$$r_\rho(t) = \rho \times s, \text{ if } \frac{d\vartheta(t)}{dt} = \frac{2\pi}{s}. \tag{17}$$

Note that, the constant reference interval shown by (17) corresponds to that used in the general SARIMA equations of (7) and (8).

Then, an ideal prediction equation of the time-variant SAR is expressed by substituting $t+h$ for t as follows.

$$\hat{y}(t+h|t) = \sum_{\rho=1}^{P} \Phi_\rho \cdot y(t+h-r_\rho(t+h)) \tag{18}$$

where the term $h - r_\rho(t+h) \leq 0$.

Figure 11 shows a schematic diagram of TVSAR model-based prediction. The reference intervals can refer the past values observed which are corresponding to the target value, on the time series with fluctuated period. Also, Figure 12 shows TVSAR model-based prediction system.

Note that, in (18), we have to know the reference interval at h sample ahead future, but it is unknown in practice. In this case, we need to estimate it. The estimation method will be explained in the next section.

3.2.1. *Online estimation of reference intervals: Short-time correlation analysis*

In TVSAR model, the reference interval $r_\rho(t)$ is an important factor to predict the lung tumor motion accurately, and must be estimated on-line. In this study, a correlation analysis is adopted to estimate the reference interval.

The correlation analysis based estimation procedure of the reference interval is as follows.

1. Calculate a correlation function between the latest subset

$$\{y(t-w), y(t-w+1), \ldots, y(t-1), y(t)\} \tag{19}$$

and k-sample lagged subset

$$\{y(t-k-w), y(t-k-w+1), \ldots, y(t-k-1), y(t-k)\} \tag{20}$$

where w is a window length for subset time series. The correlation function is given by

$$\gamma(t,k) = \frac{1}{w} \sum_{j=0}^{w-1} \frac{y(t-j) - \mu_t}{\sigma_t} \frac{y(t-k-j) - \mu_{t-k}}{\sigma_{t-k}} \tag{21}$$

Here μ_t and σ_t are the sample mean and standard deviation of the subset time series.

2. The estimated reference intervals $\hat{r}_\rho(t|t)$ can be obtained by the intervals between $k = 0$ and the peak points of the correlation function $\gamma(t,k)$ corresponded to each seasonal order ρ:

$$\hat{r}_\rho(t) = \arg \max_{r_\rho(t-1)-l \le k \le r_\rho(t-1)+l} \gamma(t,k) \tag{22}$$

where l defines the search area for the peak of $\gamma(t,k)$.

3. The window length is updated at each time by using the latest estimation of the first order reference interval as

$$w = \text{round}(0.5\hat{r}_1(t|t)). \tag{23}$$

Note that the windows length can affect the accuracy and response time to estimate the fluctuated periodicity. In this study, the window length was empirically set as almost half of single wave.

The initial condition is defined as follows.

$$\hat{r}_\rho(1|1) = \rho \times \bar{s}, \tag{24}$$

$$w = \text{round}(0.5\hat{r}_1(1|1)) \tag{25}$$

where $\bar{s}$ is the average of pre-observed time variant periods of the time series. $\bar{s} = 90$ sample was used for this study.

An example of the correlation function and estimated reference intervals are shown in Figure 13. The reference intervals are thus estimated as intervals from lag 0 to ρ-th local maximum points of the correlation function.

Time-variation of the correlation function and estimated reference intervals are shown in Figure 14. As is clear from this figure, reference intervals of lung tumor motion intricately change with time evolution.

We can now estimate the reference intervals, but we need future value of them for (18). According to the figure 14, these reference intervals also fluctuate intricately and those prediction is difficult. So $r_\rho(t+h)$ cannot be directly used for the prediction in (18). As a realistic way, we simply extrapolated $r_\rho(t+h)$ from the current estimation $\hat{r}_\rho(t|t)$ with zero-order hold: $\hat{r}_\rho(t+h|t) = \hat{r}_\rho(t|t)$. Then, (18) can be rewritten as

$$\hat{y}(t+h|t) = \sum_{\rho=1}^{P} \Phi_\rho \cdot y(t+h-\hat{r}_\rho(t|t)). \tag{26}$$

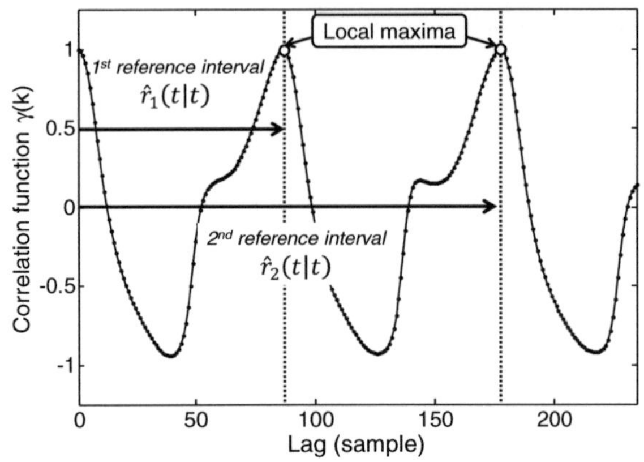

Figure 13. An example of the correlation function $\gamma(t,k)$ and estimeted reference intervals $r_\rho(t), \rho = 1$ and 2 for the tumor motion shown in Figure 5.

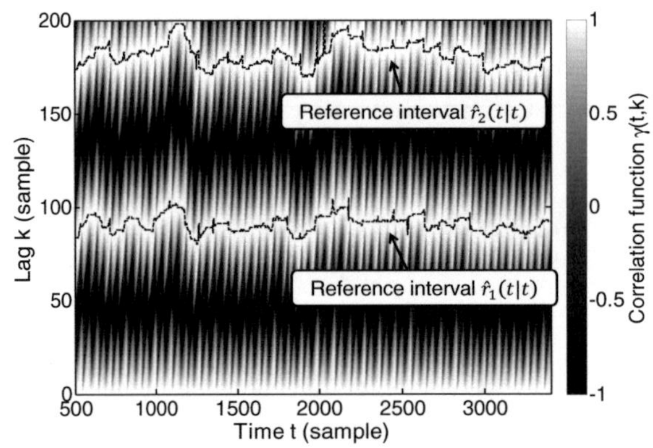

Figure 14. Time-variation of the correlation function $\gamma(t,k)$ and estimated reference intervals $r_\rho(t), \rho = 1$ and 2 for the tumor motion shown in Figure 5.

4. Experimental results

To evaluate the prediction performance of the methods described in Section 3, a prediction experiment by using clinical data sets was performed.

4.1. Experimental setup

4.1.1. Clinical data sets of lung tumor motion

Three data sets of respiratory tumor motion time series were used for the experiment. All the data sets were observed with sampling frequency $F_s = 30$ Hz and provided by Hokkaido University Hospital. Note that, a part of the data sets and measuring condition have been already shown in Section 2. Each data sets include three time series that correspond to spatial axes; left-right (LR), superior-inferior (SI), and anterior-posterior (AP) axes, respectively. The observational noises included in original data sets were preliminarily eliminated by using the statistical and low-pass filters. Figure 15 shows the three data sets and Table 1 summarizes data sets characteristics.

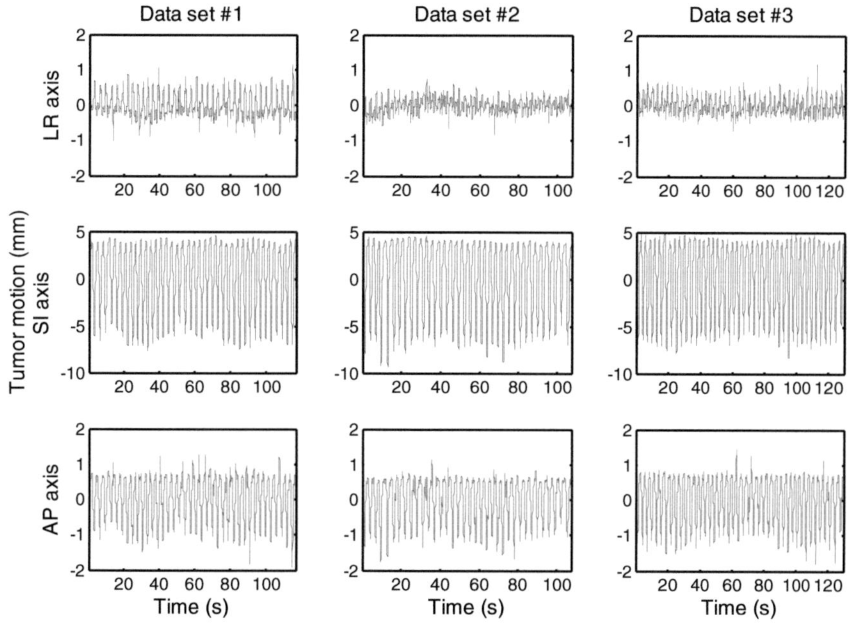

Figure 15. Three data sets of lung tumor motion. Each rows correspond to LR: left-right, SI: superior-inferior, and AP: antero-posterior axes.

4.1.2. Tested methods

For comparison of the prediction performance, the following four methods were tested.

1. Zero-order hold (ZOH) method assumes that the latest position of tumor will not change in future. The prediction equation is given by

$$\hat{y}(t+h|t) = y(t). \tag{27}$$

Note that the use of ZOH corresponds to the case that the system latency in radiotherapy machine is not compensated.

2. First-order hold (FOH) method assumes that the latest position and velocity will not change in future. The prediction equation is given by

$$\hat{y}(t+h|t) = y(t) + (y(t) - y(t-1))\,h. \tag{28}$$

3. Seasonal autoregressive (SAR) model-based prediction given in (12). Note that the time-variant period is given as $s(t+h) = \hat{r}_1(t|t)$.

Case #	Range of motion (mm)			Length	Evaluation coverage		Average period
	LR	SI	AP	T samples	t_s	t_e	$\bar{s}$
#1	0.7	10.2	1.5	3500 (116.7 s)	500 (16.7 s)	3500 (116.7 s)	91 (3.034 s)
#2	0.4	11.0	1.6	3228 (107.6 s)	501 (16.7 s)	3228 (107.6 s)	89 (2.968 s)
#3	0.5	10.9	1.6	3900 (130.0 s)	502 (16.7 s)	3900 (130.0 s)	91 (3.034 s)

Table 1. Brief summary of tested data sets. Range of motion is average distance from exhalation to inspiration.

4. Time-variant SAR (TVSAR) model-based prediction given in (26).

Parameters for SAR and TVSAR models are summarized in Table 2.

4.1.3. *Performance indexes*

To evaluate the prediction performance, mean absolute error (MAE) was calculated. MAE is given as a function of prediction horizon h as follows.

$$\mathrm{MAE}(h) = \frac{1}{t_e - t_s} \sum_{t=t_s}^{t_e} |e(t+h|t)| \tag{29}$$

Here t_s and t_e are lower and upper bounds for the evaluation shown in Table 1, and $e(t+h|t)$ is a prediction error defined by the Euclidean distance between the actual position and the predicted position.

The prediction error is calculated as follows.

$$e(t+h|t) = \sqrt{\sum_{i=1}^{3} (\hat{y}_i(t+h|t) - y_i(t+h))^2} \tag{30}$$

where $i = 1, 2$ and 3 denote LR, SI, and AP directions, respectively.

In addition to MAE, prediction success rate was adopted to evaluate the prediction efficiency for short treatment time. While treatment, the irradiation must be interrupted when unacceptable distance between the irradiated field and the target volume is detected for patient safety. Then, frequent prediction failure may prolong the treatment fraction. Thus, the rate of prediction success can discover better method. The prediction success rate (PSR) is defined as follows.

$$\mathrm{PSR}(h) = \frac{\text{Number of successfully predicted samples}}{\text{Number of all evaluated samples}} \tag{31}$$

where the numerator is calculated as the number of the prediction errors within a threshold for tolerant accuracy. In this study, the threshold was set as 1 mm.

4.2. Results

Figure 16 shows 15 samples (0.5 s) forward prediction examples of tested methods. This examples were performed on SI direction of the data set #1. According to the prediction examples, SAR and TVSAR predictions seems smooth and similar to the actual position. Their errors are around zero, but sometimes the errors become larger than 2 mm. Also, there is a

Parameters	
Order P	2
Coefficients $\Phi_\rho, \rho = 1, 2, \ldots, P$	$1/P$
Maximum lag for correlation function	400

Table 2. Parameters for SAR and TVSAR models

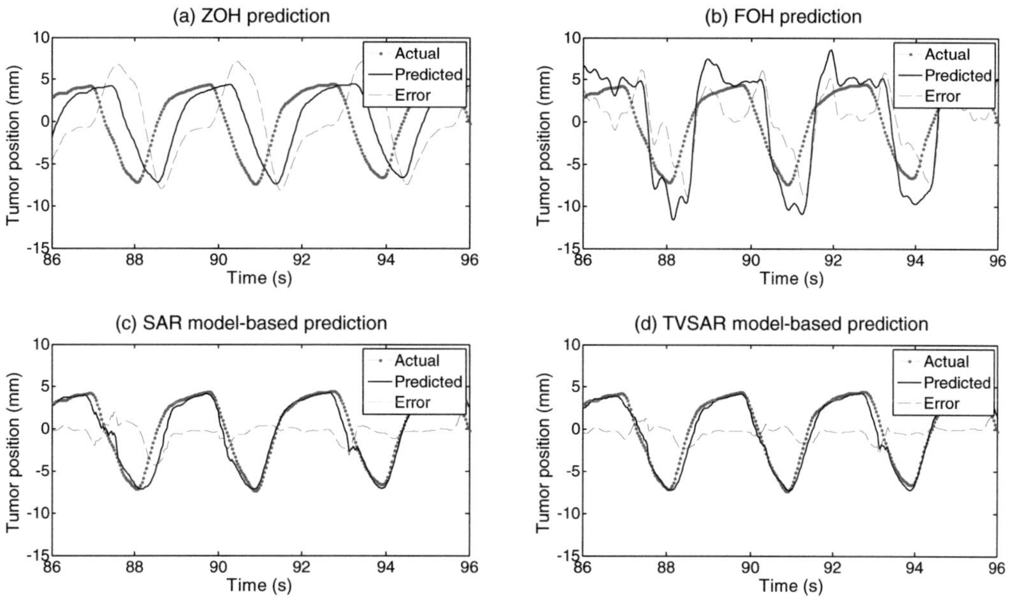

Figure 16. Examples of 0.5 s forward future prediction. (a) Zero-order hold, (b) First-order hold, (c) Seasonal autoregressive model, and (d) Time-variant seasonal autoregressive model. Gray dots, black lines, dashed lines denote actual position, predicted position, and distance between the actual and predicted positions.

difference between SAR and TVSAR predictions at 87 to 89 s. Thus, TVSAR has provided better prediction. On the other hand, ZOH and FOH predictions are noisy and not fitted into the actual position. Their errors are frequently larger than 5 mm.

Figures 17 and 18 show MAE and PSR as a function of prediction horizon h/F_s (s). Each curves of prediction performances drawn in the figures are averaged over the data sets. Also, error bars indicates standard deviation of each performances for data sets variation.

As is clear from the figures, TVSAR is superior to SAR constantly. This suggests that the basic concept of the TVSAR is more proper than SAR for the lung tumor motion prediction, i.e., reference intervals can work to improve the prediction accuracy for quasi-periodical nature. Then, MAE and PSR curves of TVSAR are the least and the highest for prediction horizon $h/F_s > 0.2$ s, respectively. It means that TVSAR is the first best method in the tested methods for the radiotherapy machine with system latency of 0.2 s or longer. The average accuracy and 70 % of the predicted samples respectively are sub-millimeter. This may suggest that the TVSAR can perform tumor following irradiation during over 70 % of a single treatment fraction.

SAR is the second-best, but the average error for prediction horizon longer than 0.5 s is over 1 mm. Note that the difference from TVSAR depends on the use of the second reference interval in this experiment. Thus, unsuitable referring the past values by using the latest period, $\rho \cdot s(t)$, increases the considerable prediction error.

ZOH and FOH are superior to SAR and TVSAR predictions for shorter prediction horizons < 0.2 s. It is not remarkable because the position and velocity don't change drastically in short term in general. The periodicity in the respiratory motion quickly changes the position

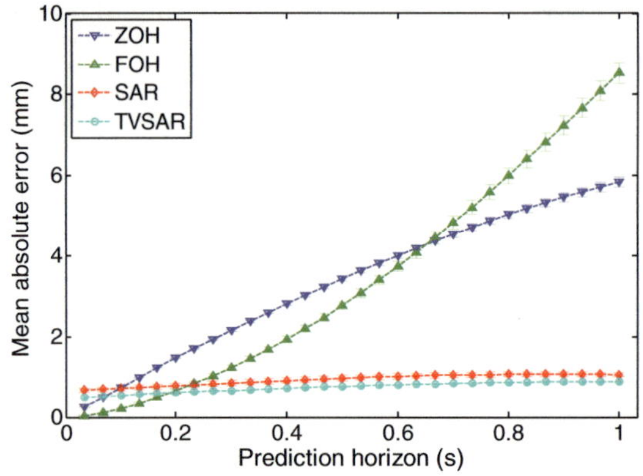

Figure 17. Mean absolute error as a function of prediction horizon h/F_s (s).

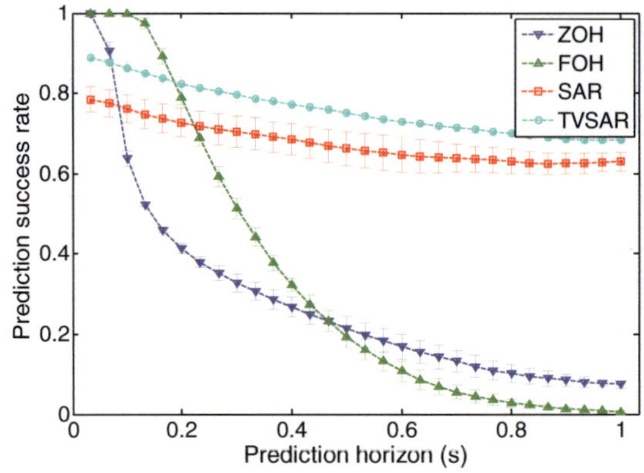

Figure 18. Prediction success rate as a function of prediction horizon h/F_s (s).

and velocity at each peaks. Then ZOH and FOH cannot take into account the changes arisen. Therefore, their prediction errors drastically increase to unacceptable level with increase in prediction horizon. Those frequent and large prediction errors cause prolong the treatment fraction.

5. Conclusion

In this chapter, a prediction method of respiratory induced tumor motion for tumor following radiotherapy were introduced. The respiratory motion involves quasi-periodic nature as a dominant component and the motion causes unacceptable error of irradiation due to the system latency between tumor localization and beam-repositioning. To compensate the latency, prediction method of respiratory motion is very important and its prediction accuracy directly affects irradiation accuracy. Then, as a potent prediction method, time-variant seasonal autoregressive (TVSAR) model was introduced. TVSAR is an extended model of seasonal autoregressive model to be adaptable to the quasi-periodical nature. An experimental result by using clinical data sets of lung tumor motion showed the average accuracy of TVSAR is less than 1 mm for up to 1 s forward prediction and TVSAR is superior to other methods tested for <0.2 s ahead future prediction. Thus, TVSAR model-based prediction method achieved accurate prediction of the respiratory tumor motion and can help the tumor following irradiation.

Acknowledgements

The authors gratefully acknowledge comments from the clinical standpoint by Dr. Yoshihiro Takai and Dr. Yuichiro Narita at Hirosaki University Graduate School of Medicine, Hirosaki, Japan. Also, the authors would like to thank Dr. Shirato and his colleagues at Hokkaido University Hospital, Sapporo, Japan. The data sets used for this chapter were observed and provided by them. This work was partially supported by Japan Society for the Promotion of Science Grant-in-Aid for JSPS Fellows Grant Number 23-8189, and Varian Medical Systems Inc., Palo alto, CA.

Author details

Kei Ichiji
Department of Electrical and Communication Engineering, Graduate School of Engineering, Tohoku University, Sendai, Japan.
Research fellow of Japan Society for the Promotion of Science (JSPS).

Noriyasu Homma and Makoto Yoshizawa
Research division of Advanced Information Technology, Cyberscience center, Tohoku University, Sendai, Japan.

Masao Sakai
Center for Information Technology in Education, Tohoku University, Sendai, Japan.

Makoto Abe
Department of Electrical and Communication Engineering, Graduate School of Engineering, Tohoku University, Sendai, Japan.

Norihiro Sugita
Department of Management Science and Technology, Graduate School of Engineering, Tohoku University, Sendai, Japan.

6. References

[1] Onishi H, Araki T, Shirato H, Nagata Y, Hiraoka M, Gomi K, et al. Stereotactic hypofractionated high-dose irradiation for stage I nonsmall cell lung carcinoma: clinical outcomes in 245 subjects in a Japanese multiinstitutional study. Cancer. 2004;101(7):1623–1631.

[2] Shirato H, Suzuki K, Sharp G, Fujita K, Onimaru R, Fujino M, et al. Speed and amplitude of lung tumor motion precisely detected in four-dimensional setup and in real-time tumor-tracking radiotherapy. International journal of radiation oncology, biology, physics. 2006;64(4):1229–1236.

[3] Suh Y, Dieterich S, Cho B, Keall P. An analysis of thoracic and abdominal tumour motion for stereotactic body radiotherapy patients. Physics in medicine and biology. 2008;53(13):3623–3640.

[4] Keall P, Mageras G, Balter J, Emery R, Forster K, Jiang S, et al. The management of respiratory motion in radiation oncology report of AAPM Task Group 76. Medical physics. 2006;33(10):3874–3900.

[5] Purdy J. Current ICRU definitions of volumes: limitations and future directions. Seminars in radiation oncology. 2004;14(1):27–40.

[6] Kubo H, Hill B. Respiration gated radiotherapy treatment: a technical study. Physics in medicine and biology. 1996;41(1):83–91.

[7] Shirato H, Shimizu S, Kitamura K, Nishioka T, Kagei K, Hashimoto S, et al. Four-dimensional treatment planning and fluoroscopic real-time tumor tracking radiotherapy for moving tumor. International journal of radiation oncology, biology, physics. 2000;48(2):435–442.

[8] Keall P, Kini V, Vedam S, Mohan R. Motion adaptive x-ray therapy: a feasibility study. Physics in medicine and biology. 2001;46(1):1–10.

[9] Poulsen P, Cho B, Sawant A, Ruan D, Keall P. Detailed analysis of latencies in image-based dynamic MLC tracking. Medical physics. 2010;37(9):4998–5005.

[10] Poulsen P, Cho B, Ruan D, Sawant A, Keall P. Dynamic multileaf collimator tracking of respiratory target motion based on a single kilovoltage imager during arc radiotherapy. International journal of radiation oncology, biology, physics. 2010;77(2):600–607.

[11] Sharp G, Jiang S, Shimizu S, Shirato H. Prediction of respiratory tumour motion for real-time image-guided radiotherapy. Physics in medicine and biology. 2004;49(3):425–440.

[12] Ma L, Herrmann C, Schilling K. Modeling and prediction of lung tumor motion for robotic assisted radiotherapy. Intelligent Robots and Systems, 2007 IROS 2007 IEEE/RSJ International Conference on. 2007;p. 189–194.

[13] Bukovsky I, Ichiji K, Homma N, Yoshizawa M, Rodriguez R. Testing potentials of dynamic quadratic neural unit for prediction of lung motion during respiration for tracking radiation therapy. Neural Networks (IJCNN), The 2010 International Joint Conference on. 2010;p. 1–6.

[14] Mizuguchi A, Demachi K, Uesaka M. Establish of the prediction system of chest skin motion with SSA method. International Journal of Applied Electromagnetics and Mechanics. 2010;33(3):1529–1533.

[15] Ruan D. Kernel density estimation-based real-time prediction for respiratory motion. Physics in medicine and biology. 2010;55(5):1311–1326.

[16] Kalet A, Sandison G, Wu H, Schmitz R. A state-based probabilistic model for tumor respiratory motion prediction. Physics in medicine and biology. 2010;55(24):7615–7631.

[17] Verma PS, Wu H, Langer MP, Das IJ, Sandison G. Survey: Real-Time Tumor Motion Prediction for Image-Guided Radiation Treatment. Computing in Science & Engineering. 2011;13(5):24–35.

[18] Box G, Jenkins G, Reinsel G. Time series analysis. Holden-day San Francisco; 1976.

[19] Homma N, Sakai M, Endo H, Mitsuya M, Takai Y, Yoshizawa M. A new motion management method for lung tumor tracking radiation therapy. WSEAS TRANSACTIONS on SYSTEMS. 2009;8(4):471–480.

[20] McCall K, Jeraj R. Dual-component model of respiratory motion based on the periodic autoregressive moving average (periodic ARMA) method. Physics in medicine and biology. 2007;52(12):3455–3466.

[21] Ichiji K, Sakai M, Homma N, Takai Y, Yoshizawa M. SU HH BRB 10: Adaptive Seasonal Autoregressive Model Based Intrafractional Lung Tumor Motion Prediction for Continuously Irradiation. Medical Physics. 2010;37.

[22] Brockwell P, Davis R. Introduction to time series and forecasting. Springer Verlag; 2002.

Chapter 6

Neutron Dose Equivalent in Tissue Due to Linacs of Clinical Use

S. Agustín Martínez Ovalle

Additional information is available at the end of the chapter

http://dx.doi.org/10.5772/56513

1. Introduction

When operating linear accelerators of clinical use at energies above 8 MeV, neutrons are produced when either electron or photon configurations are used [1]. This is mainly due to the interactions of photons and electrons of such energies with the high-Z materials present in the accelerator head (target, scattering foils, collimators, etc.) [22]. Because of their high relative biological effectiveness, photoneutrons are a particular source of unwanted out-of-field exposure of patients and several authors have pointed out the possibility of associated risks of secondary cancers after radiotherapy.

The overdose due to neutrons in patients undergoing radiotherapy is difficult to measure or estimate. Neutron fluence and spectra in water have been measured using bubble detectors and superheated drop detectors [8, 9, 21], ^{197}Au-based Bonner spheres [12] and thermoluminescent dosemeters [11, 35, 40]. [10] measured the neutron fluence at the patient plane for various linacs using gold-foil activation. [13], using the same technique, measured neutron spectra for various linacs and determined neutron fluence and ambient dose equivalents. All these measurements offer valuable information that can be compared with the results of Monte Carlo simulations.

The Monte Carlo simulation has been used to study different problems linked to neutron dosimetry. [18] calculated neutron fluence and spectra at different positions surrounding a Varian Clinac 2100C/2300C linac. [31] studied the production of neutrons in the high-Z components of a Siemens Mevatron linac. [7] investigated the field size effects, off-axis dose profiles, neutron contribution from the linac head, and dose contribution from capture gamma rays, phantom heterogeneity effects and effects of primary electron energy shift in some Linac configurations. [41] calculated neutron ambient dose equivalent for different collimator configurations in a Varian Clinac 2300 C/D. [34] studied the effects of modeling different accelerator head and room geometries on the neutron fluence and spectra for a Siemens Primus Linac. [3] determined neutron doses to critical organs for a Siemens Mevatron KDS. Different versions of the Monte Carlo N-particle transport code [5] were used in all these works. Recently [37], dose to patients due to the emitted photoneutrons were calculated by carrying out the simulations with Geant4 [2], the latest generation of the old Geometry and Tracking (GEANT) Monte Carlo code.

Even the Electron Gamma shower (EGS) Monte Carlo code, in particular EGS4 [30], was used to investigate neutron sources in a Varian Clinac [23].

Despite the large amount of calculations available, there is not much information about the increase in the dose to patients due to neutrons produced in the linac head. Only very recently, a detailed study was carried out by [19], who calculated neutron spectra and dose equivalent in tissue for a Varian Clinac.

Another problem occurring with the previous works is that neutron fluence and doses in radiotherapy were analyzed with different methodologies, for various Linacs, patient or phantom models, energies, field sizes, gantry angles, treatment modes etc. As a consequence, the results obtained until now have significant differences between them as some authors have pointed out [11, 21, 40].

The neutron contribution yielded by some linacs commonly used for radiotherapy was evaluated using the Monte Carlo code MCNPX (v. 2.5) [33]. Eight different configurations for linacs of three different manufacturers have been considered. The approach includes two main points. First, the various linacs have been analyzed using the same methodology, thus permitting a meaningful comparison between the results obtained. Secondly, we have focused on the dose to patients. Thus, we have calculated neutron fluence, neutron spectra, absorbed dose and dose equivalents in various points of an ICRU tissue phantom. The results from studies done in this chapter have led to several publications [24–26, 36].

2. ICRU tissue phantom

The phantom used in all cases, a phantom of $100 \times 50 \times 30$ cm^3, made of ICRU tissue (11 % Carbon, 76.2 % Oxygen, 10.1 % Hydrogen and 2.6 % Nitrogen in weight) [16], simulating a patient was situated with its surface at 100 cm from the source. The Fig. 1 shows schematically the phantom half, as used in the simulations.

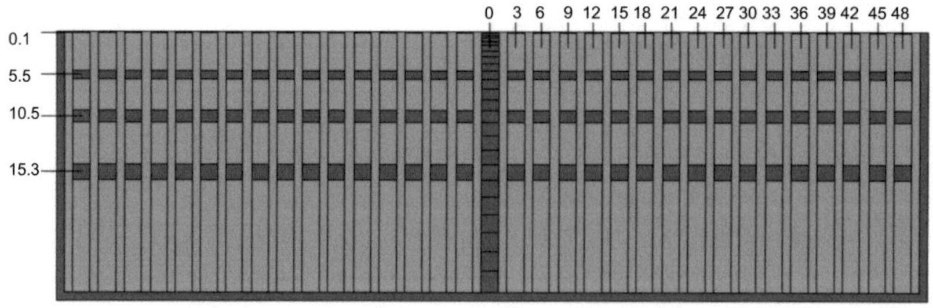

Figure 1. Outline of 1/2 ICRU phantom indicating the position of the cells in direction at axis Z radiation and outside the axis.

Dark gray cells are the positions where the neutron determinations and shown correspond to the axis and transverse positions situated off axis, between 0 and 48 cm, and 0.1, 5.5, 10.5 and 17.1 cm deep. In order to improve the statistical distribution of the cells with depth in the radiation beam axis increases proportionally with the depth from $2.0 \times 2.0 \times 0.1$ cm^3, in the surface, to $2.0 \times 2.0 \times 2.4$ cm^3, in deep.

3. Description of linacs

We studied three brands of accelerators. The simulated models correspond to the linacs Varian Clinac 2100C/D for configurations of 10, 15, 18, and 20 MV, Elekta Inor of 15 MV Elekta SL-25 18 MV and Siemens KDS 18 MV. The geometries were constructed according to manufacturer's specifications and following some recommendations from previous work. Both the jaws as the system multileaf (MLC) were fitted in all cases to achieve a treatment field of 10×10 cm^2, with which all simulations were performed.

		Siemens KDS	Elekta Inor	Elekta SL25	Varian Clinac	
		18 MV	15MV	18 MV	15 MV	18/20 MV
target	materials	Au	W/Re	W/Ni/Fe	W	
	[%]	100	90/10	95/3.75/1.25	100	
	ρ [g cm^{-3}]	19.3	19.4	18.0	19.3	
target cover	materials	Cu	Cu		Cu	
	[%]	100	100		100	
	ρ [g cm^{-3}]	8.96	8.96		8.96	
primary collimator	materials	W	W/Ni/Fe	Pb/Sb	W	
	[%]	100	95/3.75/1.25	96/4	100	
	ρ [g cm^{-3}]	19.3	18.0	11.12	19.3	
flattening filter	materials	Cr/Fe/Ni	Cr/Ni/Fe		W	Ta/Fe
	[%]	18/74/8	18/74/8		100	–
	ρ [g cm^{-3}]	8.03	8.03		19.3	16.65/7.874
secondary collimator	materials	W	W/Ni/Fe	Pb/Sb	W	
	[%]	100	95/3.75/1.25	96/4	100	
	ρ [g cm^{-3}]	19.3	18.0	11.12	19.3	
multileaf collimator	materials		W/Ni/Fe	Pb/Sb	W	
	[%]		95/3.75/1.25	96/4	100	
	ρ [g cm^{-3}]		18.0	11.12	19.3	
jaws	materials	W	W/Ni/Fe	Pb/Sb	W	
	[%]	100	95/3.75/1.25	96/4	100	
	ρ [g cm^{-3}]	19.3	18.0	11.12	19.3	

Table 1. Materials of the various elements of the linac heads considered in this work. The percentage compositions and densities, ρ, are also given. The Varian Clinac flattening filter for 18 and 20 MV is made of Ta with a cover of Fe.

The differences between marks of accelerators, are primarily concerned with the materials used in the construction of each of the parts of the head of the linacs, as the target for X-ray production, which is usually embedded within a shell material which is usually Cu, the flattening filter or filters, with some models of those studied here compose of a double filter flatter built of different materials depending on the marks, the MLC and the jaws, which are usually constructed of W, and the outer shield , which is usually of Pb and Fe. All these components, are responsible of the production of neutrons in the head of the accelerator. The materials that make up each of these elements and their densities for the accelerators studied here are summarized in Table 1.

The evaluation of this production cannot be neglected if one considers that these devices produce neutrons to the order of 1012 neutrons per Gy in conventional radiotherapy treatment [10]. Furthermore it has been demonstrated in previous studies [23]; [34]; [22] that in fact, neutron production is defined by the material used and its respective threshold to the photonuclear reactions. For example, it is noted that Varian models used materials (see Table

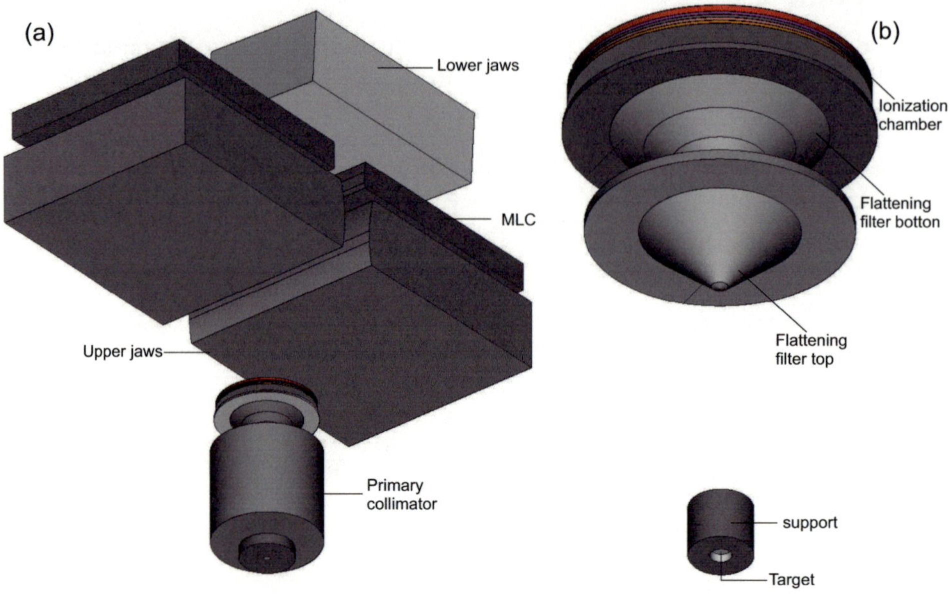

Figure 2. Geometry of an Elekta accelerators.

1), with Z higher than those used by other models for the construction of the head. This aspect will be reflected in the production of neutrons of the Linacs.

The mechanism that allows the generation of neutrons, is mainly the reaction (γ, n), which depends on the energy of the photons incident on these materials. The threshold energy of the incident photon to produce this type of reaction is 7.6, 6.2, 6.7, 13.1, 7.6 and 8.1 MeV for Ta, W, Pb, Al, Fe and Au, respectively [27], which are the main components of the target, flattening filter, jaws, MLC systems and shield of different teams studied. This means that the photoproduction mechanism is fully guaranteed for accelerators with energies above 10 MV.

Fig. 2a shows the geometry of an Elekta accelerator. These geometries correspond the Inor model of 15 MV and SL25 of 18 MV, in this case, the geometries of the two models have the same characteristics in terms of dimensions of the various elements; the differences are in the materials used in construction of the target, MLC system and jaws as seen in Table 1. Fig. 2b shows the target (1 mm in diameter) and target cover of Cu with the dual flattening filter system, constructed of stainless steel.

Each of the accelerators studied requires a previous tuning process, in order to establish the energy of incident electrons, to make it suitable for calculations. This process is carried out by comparison between the curves of percentage depth dose (PDD) simulated and measured, the latter provided by each of the radiophysics services from hospitals in which the respective model is studied. However, for models Varian Clinac 2100 C/D of 10 MV and 20 MV, it was not possible to get the experimental PDD.

The process starts by estimating the energy values around the nominal energy value of the accelerator. For each of these energies: $TPR_{20,10}$ maguitude is calculated. Which is the amount recommended by dosimetry protocols based on both air kerma patterns, as in patterns of absorbed dose in water [14, 17, 39]. This quantity is defined as the ratio between the absorbed

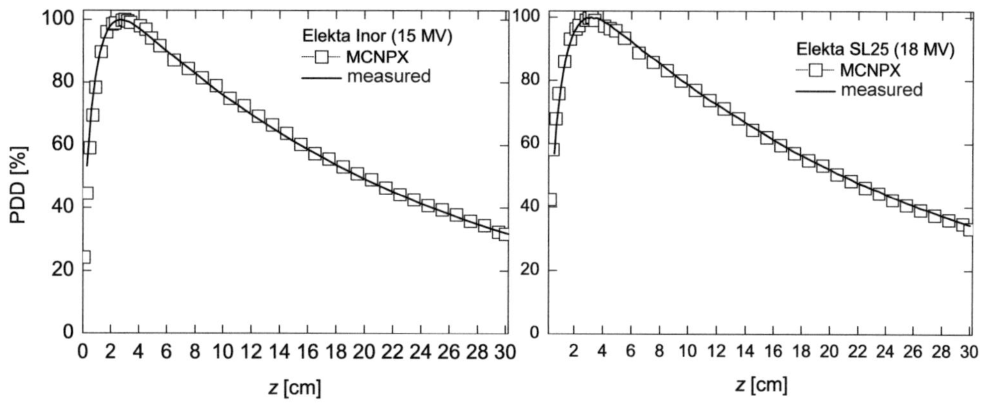

Figure 3. Comparison of PDD's calculated with MCNPX for Inor Elekta accelerators of 15 MV (left) and Elekta SL 25 of 18 MV (right) with those measured experimentally.

dose in the beam axis at 20 cm and 10 cm deep in a water phantom, obtained with a source detector distance constant of 100 cm and a field size 10×10 cm^2 in the position of the detector.

Having determined the values of $TPR_{20,10}$ to the energies considered, a calibration curve is established and from the experimental value of $TPR_{20,10}$ a tuning energy is obtained. Fig. 3 compares the experimental PDD, that were measured in hospitals Ramón and Cajal of Madrid (Spain) and Hospital Río Hortega of Valladolid (Spain), with those obtained after tuning to Inor Elekta accelerators of 15 MV and 18 MV SL-25. As we see the agreement is excellent.

The value of the maximum dose due to photons supplied by the accelerator in the build-up region is extacted from the simulated PDD. In the case of Elekta accelerators, these maximum values are 6.06×10^{-16} y 1.09×10^{-15} Gy of photons emitted per electron, and are at 3.0 and 3.2 cm depth, respectively. These depths are in good agreement with those published in the [4], for accelerators of this energy. The dose values found in the build-up region are the value reference against which the dose equivalent due to photoneutrons is expessed.

Let's say in conclusion that for Elekta models, the electron beam incident on the target is simulated by a Gaussian of mean value 13.77 MeV and 0.8 MeV of FWHM for the Elekta Inor, and 16.1 MeV and 1.5 MeV for the Elekta SL-25.

The next accelerator that was studied is the Siemens Mevatron KDS in configuration of 18 MV (Fig. 4), the electron beam that impinges on the target (also of 1 mm diameter) was simulated using a Gaussian of average value of 15.5 MeV and 1.5 MeV of FWHM. This energy is selected, Based on the previous tuning of the accelerator by means of the experimental PDD and was provided by the Hospital Universitary St. Cecilio of Granada (Spain) (Fig. 5). The maximum dose due to photons supplied in the region of the build-up is de 4.48×10^{-15} Gy of photons per emitted electron and is 3.2 cm deep.

The last accelerator studied and one of the most commonly found with dual energies of 6 and 15 MV or 6 MV and 18 MV, is the accelerator Varian Clinac 2100 C/D. Fig. 6 shows the geometry corresponding to the configurations of 10, 15, 18 and 20 MV of photons. In this case the electron beam incident on the target of 1 mm diameter is simulated, respectively, by a monodirectional and monoenergetic beam of 10.5, 15.04, 18.3 and 20.5 MeV. These energies are obtained in tuning of each of the accelerators, by comparison with experimental PDDs,

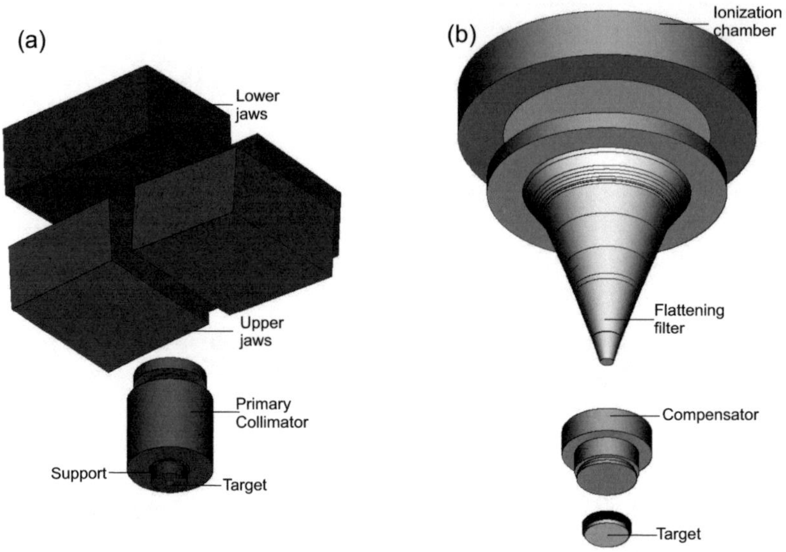

Figure 4. Geometry accelerator Siemens KDS of 18 MV.

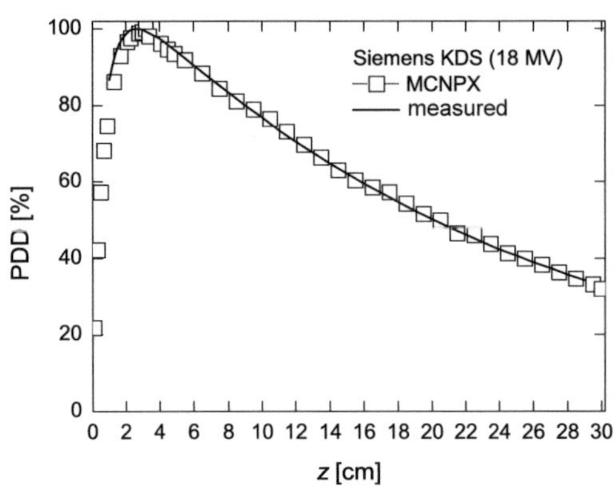

Figure 5. Comparison of PDD calculated with MCNPX for accelerator Siemens KDS of 18 MV and measured experimentally.

which were supplied by the Hospital Clinic Universitary of Valladolid (Spain) (for 15MV) and the Hospital Virgin of the Nieves in Granada (Spain) (for 18 MV).

The energies of tuning for the accelerator 10 and 20 MV were taken from [23], who simulated these models. The maximum dose due to photons in these accelerators in the build-up region are 2.78×10^{-16}, 4.85×10^{-16}, 6.57×10^{-16} and 1.18×10^{-15} Gy of photons per emitted electron, and are at 2.8, 3.0, 3.2 and 3.4 cm depth, respectively, for the four configurations analyzed.

In Fig. 6a, the geometry maintains the same dimensions for the four configurations studied, except for materials used in the manufacture of components (see Table 1). Fig. 6b shows, the target and your cover and the flattening filter. These accelerators are equipped with MLC

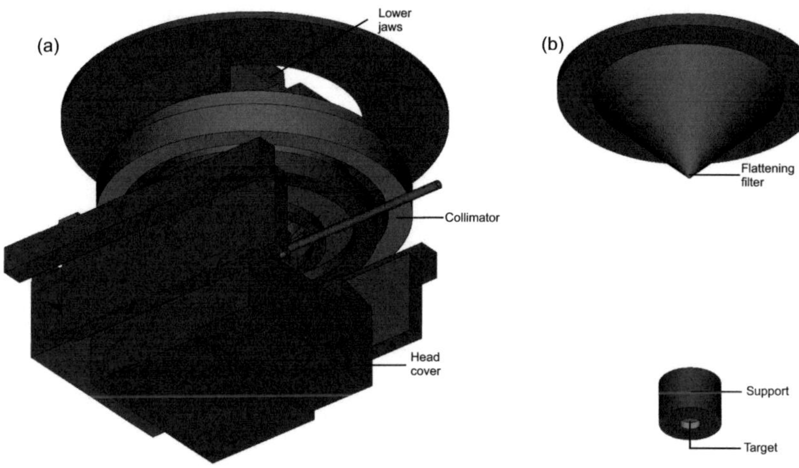

Figure 6. Geometry the accelerators Varian Clinac 2100 C/D.

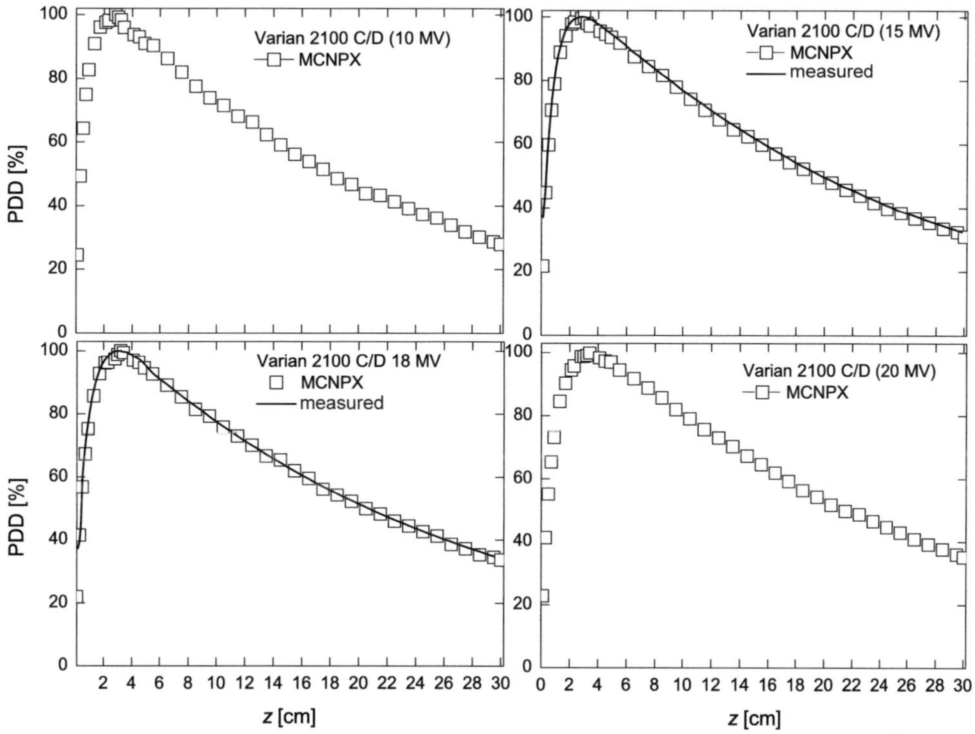

Figure 7. Comparison of PDD calculated with MCNPX for accelerators Varian Clinac 2100 C/D of 10, 15, 18, and 20 MV and the measured experimentally for 15 and 18 MV.

system, similar to the Elekta models studied. Fig. 7 shows the different PDD's of tuning for each Varian accelerators studied

In this case, the geometry used for the simulations was constructed from 93 geometric elements which include cones, cylinders, spheres, truncated cones, parallelepiped, triangular prisms, pyramids triangular, and quadrangular pyramids. In the simulations the Elekta accelerators, the shielding cover of Varian accelerator was included, because it was not possible to obtain data on the dimensions of the shielding cover of Elekta accelerator, considering that the materials used by different manufacturing for shielding the head are similar. wherever the shielding cover is not included we have called it geometry simplified in contradistinction is full geometry where the shielding cover is included.

The objective to study simplified and full geometries is due to the different positions opposing about the shielding against neutron radiation. Works like of the [23] established minimal differences in using of simplified geometries and complete for calculating photoneutrons. This has led most authors to use simplified geometries in their calculations [32]; [20]; [7]; [42]; [34]; [3]; [28]. Taking advantage that we have the complete geometry of the accelerator Varian Clinac 2100 C/D, we have studied the influence of the shielding cover in this study.

4. Monte Carlo simulation

As noted above, once the tuning of each accelerator, was held from the simulated PDD, the depth at which it obtain the maximum absorbed dose due to photons in the region of the build-up was determinated. This value is used as reference to express all the calculations reported here for neutron dosimetry.

4.1. Neutron fluence in ICRU tissue phantom

The fluence of neutron in each accelerator is initially calculated as a function of depth into the phantom and on the axis of central radiation.

Fig 8 shows the results of fluence for the six accelerators studied. The white squares correspond to the Varian accelerators, the black circles to the Elekta accelerators and the white circles to the Siemens accelerators. The results have been grouped in each panel according to energy. As seen, the shape of the fluence curve as a function of depth on the phantom is similar in all accelerators. The differences between them are summarized in Table 2, showing the highest values of fluence, Φ_{max} and depth at which this maximum is reached, d_{max} . As we see, this depth varies between 2.03 and 2.55 cm in all cases except for the Elekta SL-25 of 18 MV for which the maximum is at 3.61 cm.

linac	Φ_{max} [cm^{-2}]	d_{max} [cm]
Siemens KDS 18 MV	$0.37 \cdot 10^{-8}$	2.20
Elekta Inor 15 MV	$0.39 \cdot 10^{-8}$	2.55
Elekta SL25 18 MV	$2.07 \cdot 10^{-8}$	3.61
Varian Clinac 15 MV	$1.46 \cdot 10^{-8}$	2.32
Varian Clinac 18 MV	$3.25 \cdot 10^{-8}$	2.03
Varian Clinac 20 MV	$5.30 \cdot 10^{-8}$	2.22

Table 2. Maximum fluence per emitted electron, Φ_{max}, and depth at which this maximum is reached, d_{max}, for the various configurations and linacs studied.

The maximum fluence of neutron, increases with the energy in accelerators of the same model. The maximum fluence is seen in the Varian 20 MV and is 1.6 times greater than in the Varian 18

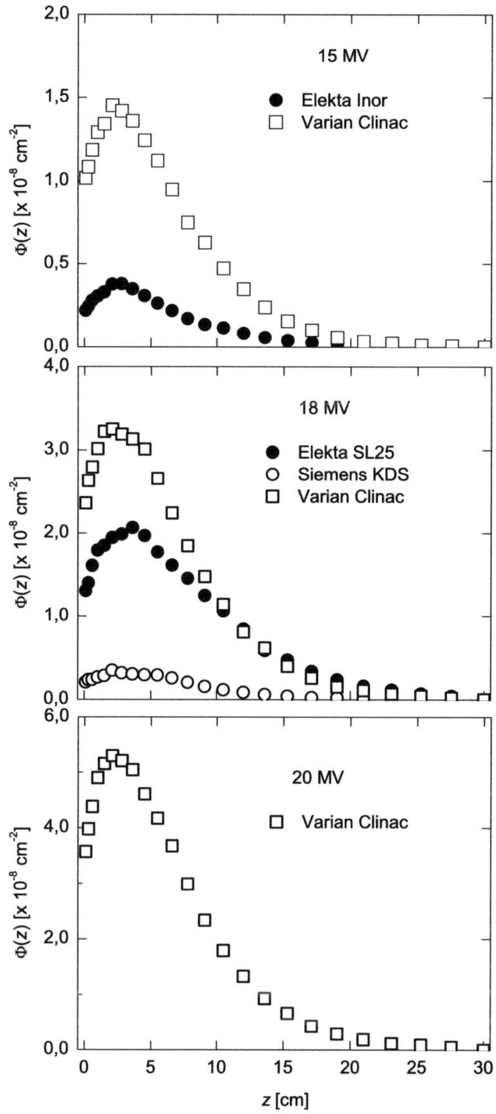

Figure 8. Neutron fluence by electron emitted as a function of depth in phantom of ICRU tissue for each of the accelerators. $Z = 0$ corresponds to the surface of the phantom.

MV, while that in Varian 18 MV it is 2.2 times greater than in the Varian 15 MV. The difference is even greater among Elekta, in which the maximum fluence of SL-25 is 5.3 times greater than that of Elekta Inor. On the other hand, the models Varian Clinac models show a considerably higher fluence than Elekta and Siemens in models of the same energy. In fact, the maximum fluence in Varian Clinac of 18 MV is 1.6 times larger than in Elekta SL-25 of 18 MV and 5.6 times higher than that found in the Siemens KDS of 18 MV. For energies of 15 MV, the maximum fluence found in the Varian Clinac is 3.7 times that found for the Elekta Inor.

The production of neutrons due to photons in each accelerator appears to be linked to the target materials and to a lesser extent, to the other elements in head. As shown in Table 1, the

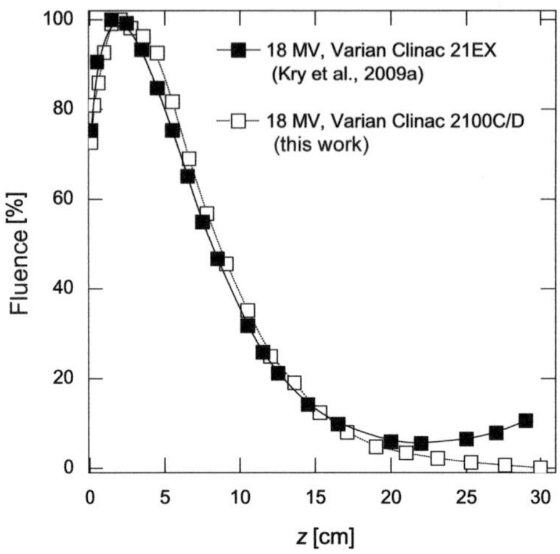

Figure 9. Comparison of the neutron fluence in a Varian Clinac 21EX accelerator of 18 MV [19] and the obtained for a Varian Clinac accelerator 2100 C/D, in this study.

target in the case of Varian Clinac is W, in the Siemens KDS is Au, while in Elekta is a mixture of W with other lighter elements. Moreover, the flattening filter in both the KDS as in Elekta models are made of lighter materials that filters of Varian. The jaws and multileaf include heavy materials in all the analyzed linacs.

Neutrons emitted from the target have average energies between 1 and 1.5 MeV. At these energies, the probability of neutron capture is negligible and neutrons mainly suffer elastic collisions (with Hydrogen), losing energy until they become thermal neutrons. Because of this, there is an increase of fluence to a maximun peak in the depth between 2-3 cm. When this maximum is reached, the neutron spectrum is more thermalized and begin to disappear by neutron capture processes, $^{14}N(n,p)^{14}C$ and $^{1}H(n,\gamma)^{2}H$, much more likely at thermal energies, and the fluence decreases monotonically with depth. It can be considered that at $2-3$cm deep the net fluence decreases moderately and with a thickness of 7 and 10cm There is radiation to half, but there is also a certain dependence of the size of the phantom.

Fig. 9 compares the fluence for two different models of accelerators of the same mark, operated at 18 MV. On one side are the results of [19] for an accelerator Varian Clinac 21EX (black squares). With the results we have obtained here for the Varian Clinac 2100 C/D. Fluence values are normalized with respect to the maximum in both the cases.

The phantom used in the two calculations is ICRU tissue, of the same geometric characteristics, and we find, an agreement until 17 cm depth, in deeper points, near the base of the phantom, we observe an increase in the curve of [19]. This may be because their simulation includes the treatment table, and it can produce backscattering of neutrons in it, which could contribute to the total fluence in the deepest zone of the phantom.

4.2. ICRU phantom in front of a neutron source

Then we studied the behavior of the ICRU phantom in front of the neutron flux coming from the linacs. The effect can be studied if we calculate the spectra of fluence of neutrons just before the phantom and inside. A tally detector was located in air at 10 cm from the surface of

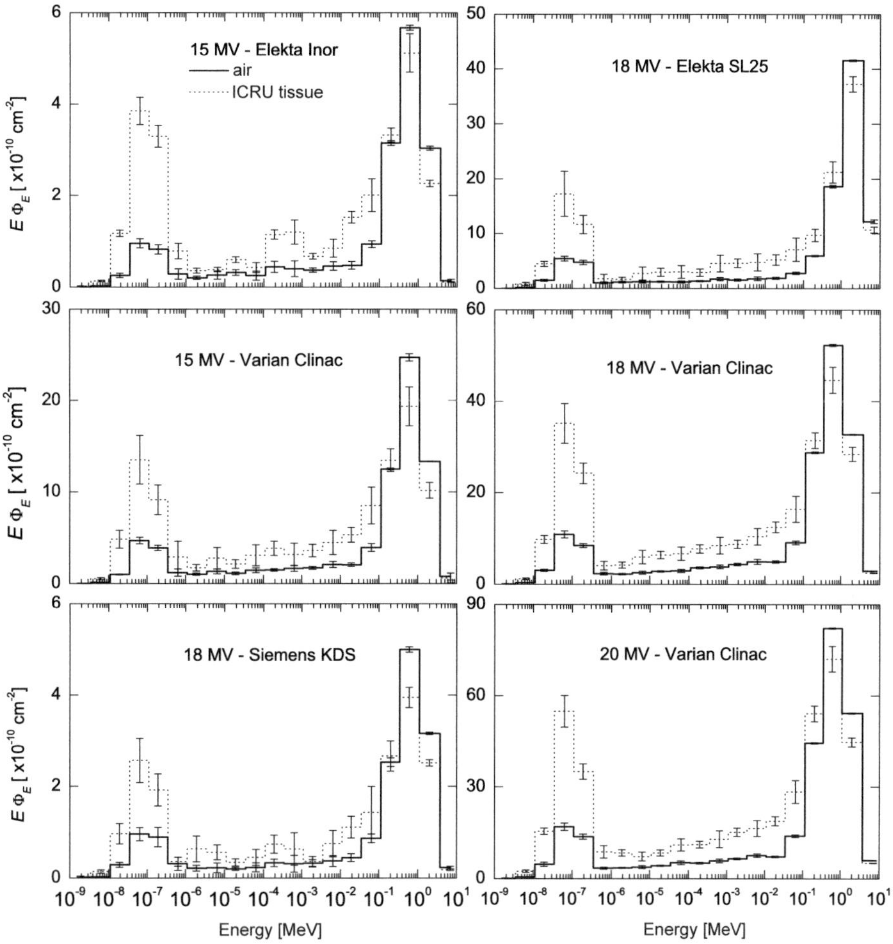

Figure 10. Spectrum of neutrons in air at 1 cm above the phantom (solid lines) and 1 cm inside the phantom (dotted lines).

the phantom, on the beam axis, and another at 1 cm into the phantom, also on the beam axis to achieve this end.

Energy spectra are shown in Fig. 10, wherein the histograms with solid lines correspond to the spectrum in air, while the histograms with dotted lines correspond to the spectrum within the ICRU tissue. We highlight some important aspects. It is observed in all cases that the peak is more pronounced at high energy and corresponds to the spectrum of fluence in air. This peak is due to fast neutrons that are emitted directly from the head of the linac.

The energy of these fast neutrons is within the energy range between 0.1 and 2 MeV proposed by [29] and [15] except the Elekta accelerator SL-25 of 18 MV, which emits neutron with energy significantly higher than the other accelerators studied.

It was also observed that at low energy, the thermal peak in tissue is greater than the thermal peak in air in all cases. This behavior is due to increased thermal neutrons that are caused by the interaction of fast neutrons primarily with the hydrogen the phantom. The thermal energy range can be considered, according to Fig. 10, between 19 meV and 0.28 eV for all accelerators studied. Between the two peaks of neutron (thermal and fast), is find an epithermal neutrons spectrum that should not be neglected in the process of thermalization.

It is generally observed that, when passing from air into phantom, a significant decrease of fast neutrons and the increase in all cases of thermal neutron and epithermal neutrons accurs in all accelerators. It is concluded that phantom behaves as a moderador of neutrons due to its high content of ^{1}H, just 11.1% of total the ICRU phantom, with a threshold of thermal production of only 2.2 MeV [8].

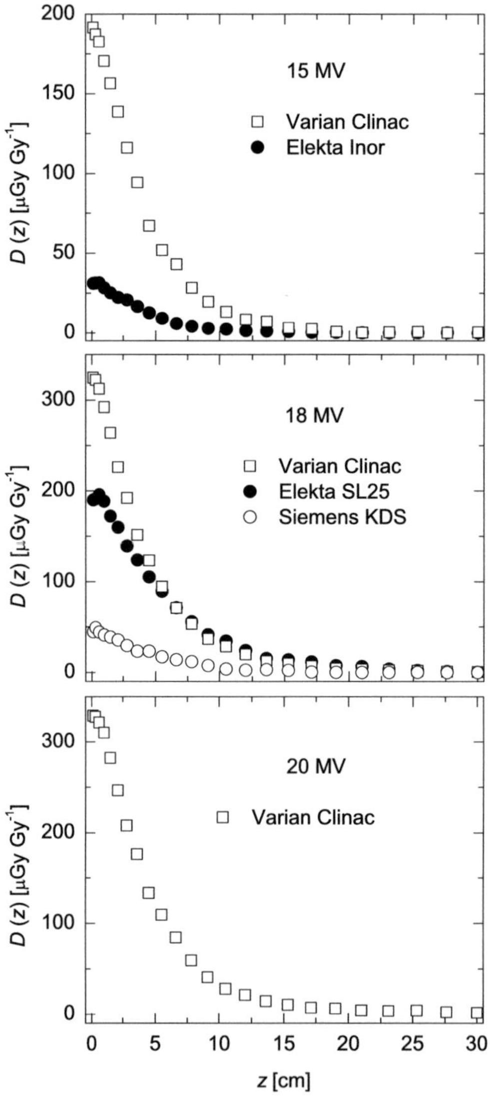

Figure 11. Absorbed dose due to neutrons as a function of depth in tissue ICRU for the different accelerators. The values are normalized to the maximum absorbed dose due to photons.

4.3. Absorbed dose due to neutrons

The radiation absorbed by the phantom ICRU, was calculated the kerma, which estimates the radiation absorbed by the medium, when exist balance of charged particles; in forward called absorbed dose, D.

The D due to neutrons was determined for all accelerators. The values obtained for the various configurations are shown in Fig. 11. The absorbed dose due to the neutrons in each case, is normalized to the maximum absorbed dose due to photons, however, we note that the absorbed dose in the Varian Clinac accelerator of 18 MV is slightly higher versus the Varian Clinac of 20 MV, although the flow is higher for the latter. This is because the values of absorbed dose due to neutrons are normalized to the maximum absorbed dose due to photons within the phantom.

The surface of phantom in Fig. 11, simulates the skin. The absorbed dose in each configuration is different according to this consideration. In the case of Varian of 15 MV this is 6.2 times greater than for Elekta Inor. The absorbed dose for Varian at 18 MV is 1.7 times greater than in the Elekta SL-25 and 7 times greater than the provided by Siemens KDS. The absorbed dose decreases with depth in all cases. To characterize this reduction, we determined depth in the tissue, at which the absorbed dose is reduced by 10% compared with the value in skin. For the Varian model, this depth is $\sim$ 9.5 cm, for Elekta models: $\sim$ 8.5 cm in the Inor linac and $\sim$ 12.5 cm for the SL-25. Finally, for the Siemens KDS it was found that this distance is $\sim$ 10 cm. These differences so marked are related, as we have been arguing, with high Z materials, used in the construction of equipment components, resulting in the generation of greater or lesser number of photonuclear reactions.

The absorbed dose decreases exponentially in the first 15 cm, in all cases, confirming that in a conventional treatment with a linear accelerator, the organs that are closer to the surface will receive a higher dose that the deeper organs.

4.4. Dose equivalent due to neutrons

The amount of radiation absorbed into tissue or organs may cause very different biological effects, and depends on the type of radiation or agent that produces particles, the value of the absorbed dose D is typically multiplied by the quality factors associated at the type of radiation, to find an equivalent in energy absorbed. The resulting quantity is called dose equivalent, H.

The dose equivalent is the amount that actually determines the biological damage to tissue or organs. The calculations of H are expressed in μSv$\cdot$UM^{-1}, relative on the maximum absorbed dose due to photons that provides each accelerator in the region the build-up and then converted to Monitor Units (MU), where 1 MU = 1 cGy. According to [19], we have:

$$H = \sum_E D(E) \cdot Q_n(E), \tag{1}$$

where $D(E)$, is the absorbed dose in tissue due to neutrons and calculated from the F6 tally ([33]), and $Q_n(E)$ is the quality factor for neutrons of energy E in the corresponding material medium, in our case ICRU tissue [38]. H can also be calculated from the spectrum of fluence as:

$$H = \sum_E \Phi(E) \cdot k(E) \cdot Q_n(E), \tag{2}$$

Where $\Phi(E)$ is the neutron fluence, calculated by F4 tally ([33]), $k(E)$ is the kerma factor for neutrons of energy E in tissue, calculated by [6], and $Q_n(E)$ arequality factors for H, C, N, and O, calculated by [38]. The product $\Phi(E) \cdot k(E)$ represents the absorbed dose $D(E)$.

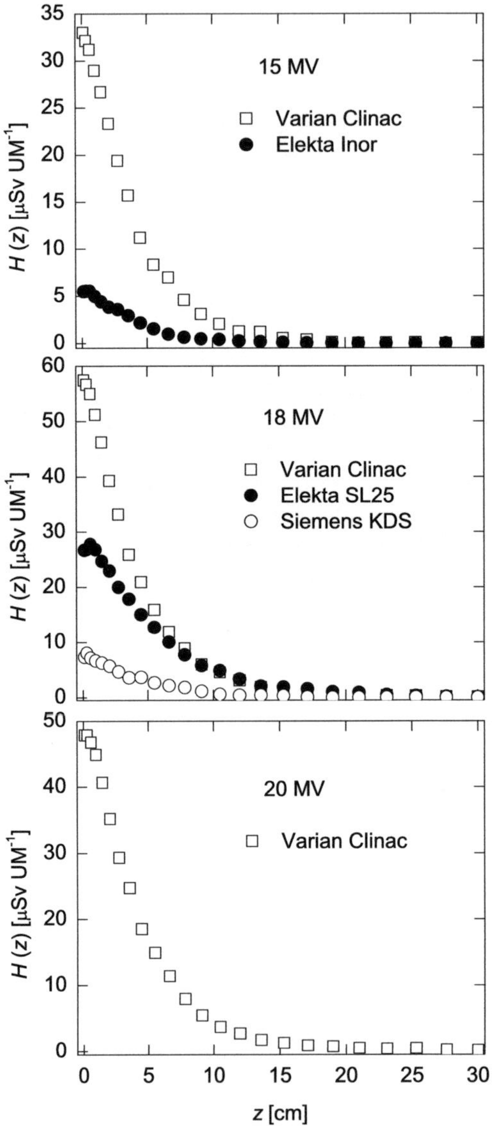

Figure 12. Dose equivalent per unit monitor in function of depth along the direction of the incident beam.

Fig. 12, shows the values of dose equivalent calculated for all accelerators studied, calculated with equation 1. If we consider the cell surface dimensions $2.0 \times 2.0 \times 0.1$ cm^3, which simulate the skin of thickness 0.1 cm, we can calculate the dose equivalent would receive this small portion of tissue, due to each accelerator which is 5.49, 7.34, 26.71, 33.01, 47.8 and 57.4 μSv$\cdot$UM^{-1} for Elekta Inor, Siemens KDS, Elekta SL25 Varian Clinac de 15, 20 and 18 MV, respectively. As can be seen the highest value corresponds to the accelerator Varian Clinac 18

MV and the lowest for Elekta Inor accelerator followed Siemens KDS. It is further noted that $H(z)$ decreases significantly after 20 cm depth in all cases.

	$H(0)$ $[\mu Gy \cdot (MU)^{-1}]$	λ $[cm^{-1}]$	χ^2 per d.o.f.
Varian Clinac 15 MV	37.0±0.9	0.263±0.007	1.45
Elekta Inor 15 MV	6.3±0.2	0.26±0.01	1.18
Varian Clinac 18 MV	64.0 ±1.0	0.251±0.004	1.52
Elekta SL25 18 MV	31.0±0.9	0.178±0.006	2.26
Siemens KDS 18 MV	8.5±0.3	0.224±0.009	0.95
Varian Clinac 20 MV	54.5±1.3	0.242± 0.006	2.81

Table 3. Values of the parameters of the fitting function 3 obtained for the dose equivalent values corresponding to the linacs analyzed in this work. The χ^2 per degree of freedom is also given.

The behavior shown by $H(z)$ depending on the depth on the phantom suggests an exponential dependence allowing a fit the data to a function of the form:

$$H(z) = H(0) \exp(-\lambda \cdot z), \tag{3}$$

where $H(0)$ is the maximum dose equivalent in surface and Z indicates the depth. The results of this adjustment are summarized in Table 3.

A first important aspect to note is the fact that the values of λ are quite similar in all cases, ranging between 0.178 and 0.263 cm$^-$1. Both $H(0)$ and λ coefficients depend on the primary spectrum of neutrons produced in first generation on the target and flattening filter of each linac by photon-neutron reactions, which in turn depends on the energy of electrons incident on the target and of course the components of the target. The coefficient λ depends of the phenomena the interaction of these primary neutrons with ICRU tissue, being predominant the elastic scattering on Hydrogen.

The cross section of elastic collision of the Hydrogen with the neutrons, depend inversely with of the energy themselves, being 0.3 cm^{-1} for 1 MeV, and between 0.1 and 0.2 cm^{-1} for 3 MeV. Note the similarity of these values with the coefficient λ which gives the fit for all the curves in Table 3.

The Fig. 13 compares the dose equivalent in depth obtained for three Varian accelerators operating at 18 MV: The Clinac 2100 C/D studied by us, the Clinac 21EX considered by [19] and a Varian generic simulated by [8] using a monoenergetic beam of neutron in direction perpendicular at phantom of water. Here the values of $H(z)$ are normalized to the maximum, H_{max}. As we see, our results (squares) are slightly on-top of the other authors in the area of intermediate depth, between 5 and 10 cm, approximately.

These differences may be related to the following motives: The accelerator studied by [19] is a Varian Clinac 21EX, equipped with a Millennium MLC of 120 sheets. In our case, the MLC is one of 80 sheets. But apart from this difference, the tuning energy used by [19] was 18.0 MeV, which differs from that set in our case (18.3 MeV). This difference in the energy of the initial electron, produce variations in the absorbed dose due to photons, and more specifically, its value at the isocenter, which, as already seen, can lead to differences in the values of H.

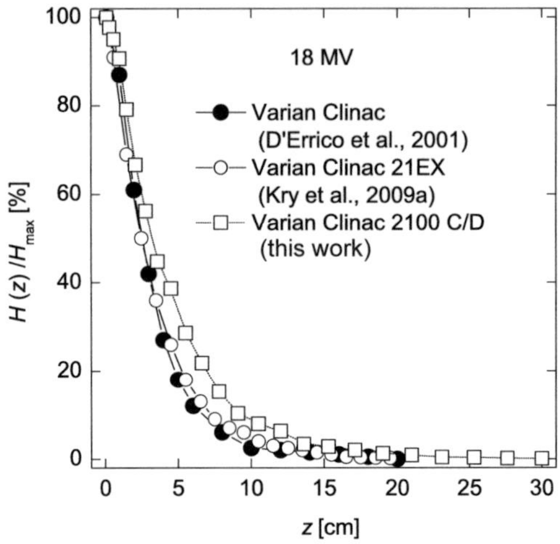

Figure 13. Comparison of the dose equivalent normalized to the maximum due to neutrons calculated in this study, with published results by [8] and [19].

The differences with the calculation of [8] are more evident due to that, as indicated; consider a monoenergetic beam of neutrons of 0.24 MeV. Both the material of phantom, and that is not considered the actual spectrum of the neutrons, are very outstanding aspects.

Now we analyze the profile of H outside the beam axis. This is important because, in a real treatment, the patient receive neutron radiation in all the body [25]. The Fig. 14 shows the results of dose equivalent profiles in function of distance from the central axis, at depths of 0.1 (open squares), 5.5 (dark squares), 10.5 (white circles) and 15.3 (black circles) cm, in the phantom ICRU for different accelerators studied.

In all cases, a general trend we observe: the dose equivalent decreases rapidly within a few inches away from the beam axis, to reach some uniformity. The region in which said reduction occurs depends on the type of accelerator, but is mainly related to the size of the radiation field. Indeed, an important part of H is transferred by neutrons with energies between 200 keV and the maximum available energy, 1-3 MeV. For distances larger than 5 cm off-axis beam (remember that the radiation field is the 10 cm^2), neutrons do not come directly to the phantom, but must pass through the jaws and in general the shield. This produces a reduction in its energy, which depend on the component materials of the head and the distance that must to traverse through within these materials. In the case of Elekta accelerators, the dose equivalent is increased at a distance exceeding 30 cm off-axis, which does not occur in other accelerators, at least as clearly. It should be noted here, that the Elekta accelerators analyzed have a flattening filter double, to which is added the fact that also a part of their secondary collimating system is oriented perpendicularly to the axis of radiation (see Fig. 2), which does not occur with the other accelerators. In conclusion the dose equivalent tends to be much more uniform and smaller, when increases the depth in the phantom. This is an important aspect to be taken into account for calculating the dose equivalent in organs [25].

4.5. Comparison of simulations and experimental measurements

If we analyze the Monte Carlo calculations and the experimental measurements, we observed differences that can be explained, taking into account the major problems when measuring

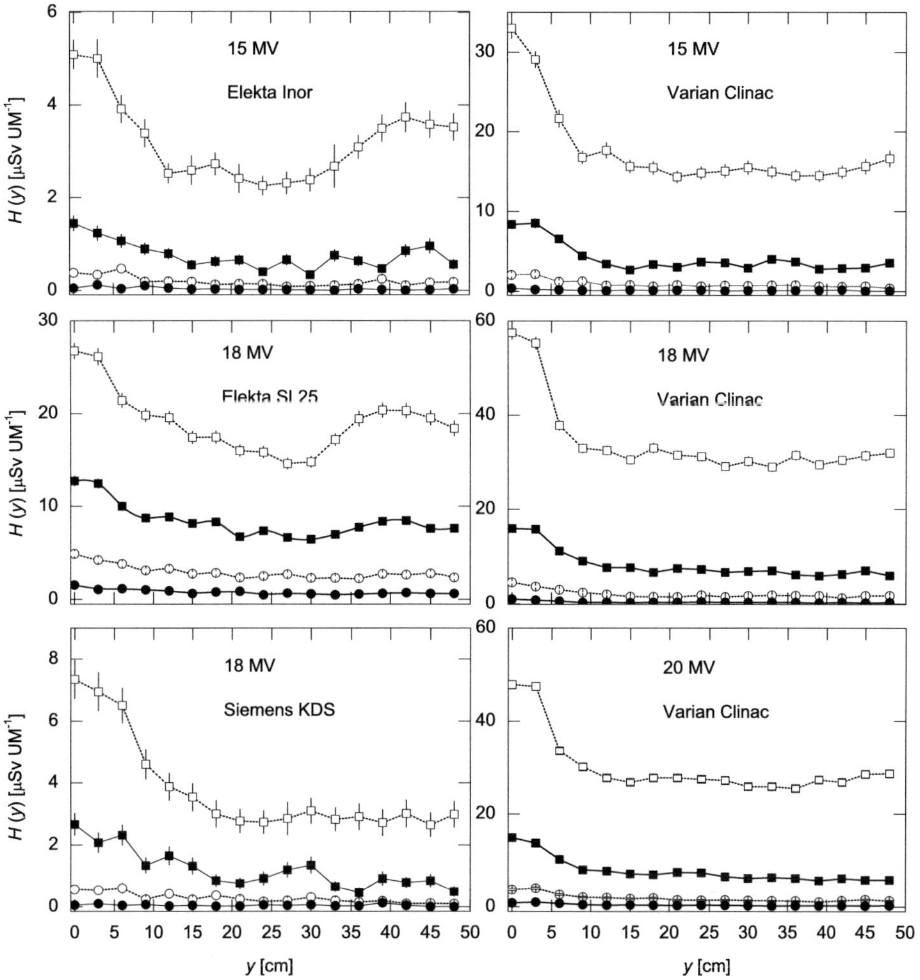

Figure 14. The profiles H to depths of 0.1 (open squares), 5.5 (dark squares), 10.5 (white circles) and 15.3 (dark circles) cm, inside the phantom, in direction transverse to the beam axis.

the neutron fluence. The sensitivity of the detectors used depends significantly on the neutron energy and this greatly influences the experimental measured of dose equivalent H.

The Fig. 15 compares experimental measurements of H, made with bubble detectors and detectors based on the CR-39 polymer by [21] and [8, 9] in water, with calculations of dose equivalent by Monte Carlo in function of depth and in the direction of central axis. The lower panel shows the results for 18 MV accelerators, together with the results obtained in our calculations for the Varian Clinac accelerators, Elekta SL 25 and Siemens KDS, in the top panel the corresponding to accelerators of 15 MV. For the Siemens Primus accelerator of 15 MV, it is observed that measures of [21] in surface are similar to the results of the calculations for the Varian Clinac of 15 MV in this work; but with increasing depth, these experimental results show much higher values than those obtained by [8, 9], which are more consistent with Monte Carlo calculations performed in this work, especially for 15 MV.

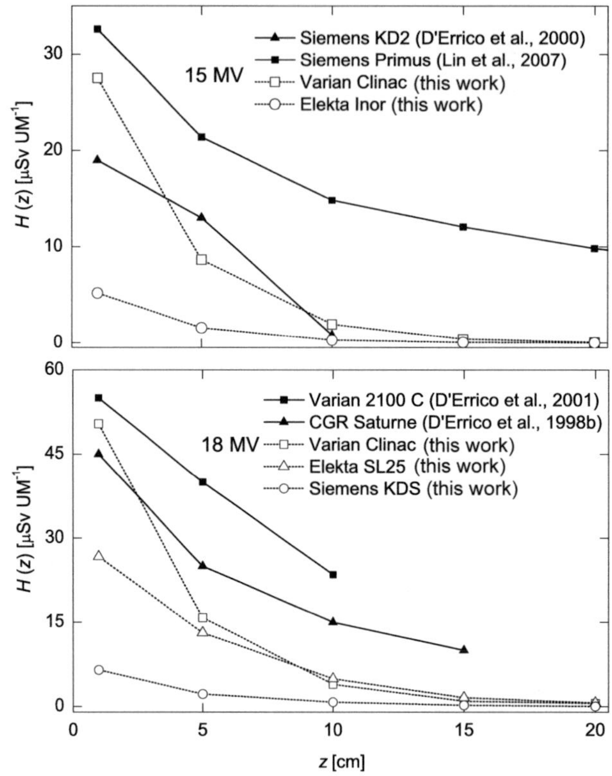

Figure 15. Dose equivalent H in function of the depth, calculated in accelerators of 15 and 18 MV, compared with measurements on a Siemens Primus of 15 MV [21], a Siemens KD-2 of 15 MV and a Varian Clinac 2100 C of 18 MV [8] and a CGR Saturne of 18 MV [9].

This is the general situation for all employees measurement systems for neutrons, without having to date, the ideal detector that can respond at any desired energy range. This is where Monte Carlo once again becomes the tool adequate for such studies.

In Fig. 16 we compared profiles dose equivalent $H(y)$ measured and calculated in this study, in function on the distance to the axis of incidence and for several accelerator of 15 MV (top panel) and 18 MV (bottom panel). The behavior of the profile of $H(y)$ is very similar in all cases, with rapid decrease in the first 10 cm, from which have small variations, except in the accelerators Siemens KDS 18 MV and Elekta Inor of 15 MV, due to low production of photoneutrons, as demonstrated above. In the top panel of Fig. 16 shows some agreement between the experimental results for the Siemens Primus accelerator and simulations for the accelerator Varian Clinac analyzed in this study.

This suggests that the bubble detectors BD-PND used have an acceptable response in surface, as mentioned before, which does not occur with BDT bubble detectors used in depth for the measurement of thermal neutrons by [21]. In general, we see that from the 10 cm off axis, $H(y)$ has small variations, very similar behavior to that found in the studied profiles in Fig. 16.

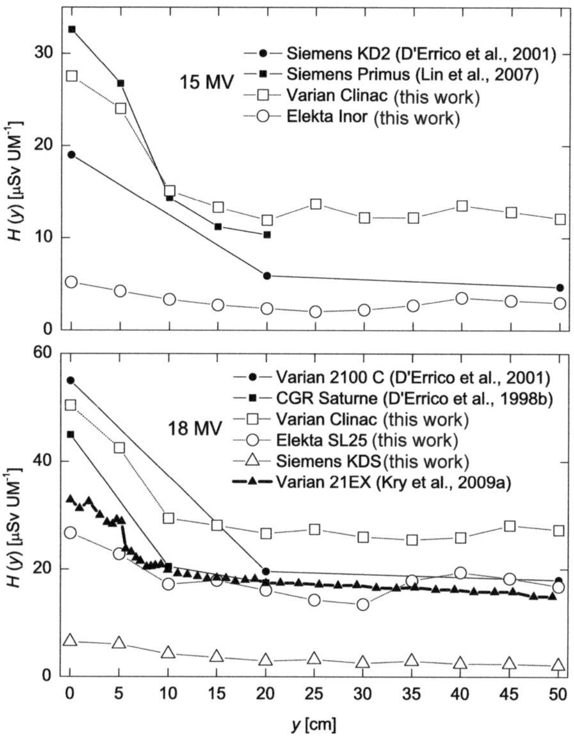

Figure 16. Profiles of dose equivalent to $Z = 1$ cm into the phantom ICRU, as function of the distance transverse to the beam axis, for linacs of 15 and 18 MV, compared with measurements of a Siemens Primus of 15 MV [21], a Siemens KD-2 of 15 MV [8], a Varian Clinac 2100 C of 18 MV [8] and a CGR Saturne of 18 MV [9] and calculations for a Varian Clinac 21EX of 18 MV [19].

5. Conclusions

In this work, the photoneutron production in four linacs with a total of six energy configurations has been analysed using the Monte Carlo code MCNPX. A detailed simulation of the geometries of the linac head has been carried out. The aim was to study the neutron dose equivalent in patients treated with these linacs in order to compare the differences between them. The important point was that the same methodology was used in all the cases, something not done till now for some of the quantities of interest here analysed.

A first result to be pointed out concerns the considerably larger photoneutron production of the Varian Clinac 2100C in comparison with the Elekta Inor and SL-25 and the Siemens Mevatron KDS for the same energy. This larger production can be linked to the materials used to built-up the target in each specific linac. In any case, the maximum fluences observed for the various linacs show a dependence almost linear with the tuned energies of the electrons incident in the target.

Neutron spectra in air, nearby the phantom and at a depth of 1 cm in the phantom were calculated. In air, the spectra are characterised, in all cases, by a pronounced peak at high energy (200 keV to 3 MeV). This peak reduces notably inside the phantom, where a peak at thermal energies appears due to neutron moderation by the medium.

Absorbed doses and dose equivalent show a similar behaviour, as a function of the depth in the phantom: they reduce strongly with the depth. This reduction can be reproduced by means of an exponential function in which the reduction rate, obtained after the corresponding fit procedure, is closely related to the cross section for elastic neutron with Hydrogen collisions at the maximum energies present in the neutron spectra obtained in the simulations.

The trend of the dose equivalent values, as a function of the transverse axis distance to the beam axis, depends strongly on the radiation field used. For large distances to the beam axis, these doses are uniform, except for the Elekta Inor and SL-25 accelerators in which they grow up. This is linked to the material present in the secondary collimators of the linacs.

One of the main points of our work concerns the determination of the dose equivalent due to neutrons inside the phantom. This has permitted to gain insight about the dose distribution in patients submitted to radiotherapy with photon beams 15 MV. As a result, it was found that surface organs are the most affected by the overdose produced by neutrons.

Author details

S. Agustín Martínez Ovalle
Universidad Pedagógica y Tecnológica de Colombia, Group of Applied Nuclear Physics and Simulation, Colombia

6. References

[1] AAPM-19 [1986]. American Association of Physicists in Medicine, *Report 19* .
[2] Agostinelli, S. e. a. [2003]. Geant4-a simulation toolkit, *Nuclear Instruments and Methods in Physics Research A* 506: 250–303.
[3] Barquero, R., Edwards, T. M., Iniguez, M. P. & Vega-Carrillo, H. R. [2005a]. Monte Carlo simulation estimates of neutron doses to critical organs of a patient undergoing 18 MV x-ray LINAC-based radiotherapy, *Medical Physics* 32: 3579–3588.
[4] BJR-25, [1996]. Central axis depth dose data for use in radiotherapy : a survey of depth doses and related data measured in water or equivalent media / prepared by a Joint Working Party of the British Institute of Radiology and the Institution of Physics and Engineering in Medicine and Biology, *British Journal of Radiology*.
[5] Briesmeister, J. F. [2000]. MCNP-A General Monte Carlo N-Particle Transoprt Code, Version 4C., *Los Alamos National Laboratory* Report: LA–13709–M.
[6] Caswell, R. S., Coyne, J. J. & Randolph, M. L. [1980]. Kerma factors for neutron energies below 30 MeV, *Radiaton Research* 83: 217–254.
[7] Chibani, O. & Ma, C. C. [2003]. Photonuclear dose calculations for high energy photon beams from Siemens and Varian linacs, *Medical Physics* 30: 1990–2000.
[8] D'Errico, F., Luszik-Bhadra, M. & Nath, R. [2001]. Depth dose equivalent and effective energies of photoneutrons generated by 6-18 MV X-ray beams for radiotherapy, *Health Physics* 80: 4–11.
[9] D'Errico, F., Nath, R., Tana, L., Curzio, G. & Alberts, W. G. [1998]. In-phantom dosimetry and spectrometry of photoneutrons from an 18MV linear accelerator, *Medical Physics* 25: 1717–1724.
[10] Followill, D. S., Stovall, M. S., Kry, S. F. & Ibbott, G. S. [2003]. Neutron source strength measurements for Varian, Siemens, Elekta, and General Electric linear accelerators, *Journal of Applied Clinical Medical Physics* 4: 189–194.

[11] Harrison, R. M., Wilkinson, M., Shemilt, A., Rawlings, D. J., Moore, M. & Lecomber, A. R. [2006]. Organ doses from prostate radiotherapy and associated concomitant exposures, *British Journal of Radiology* 79: 487–496.

[12] Howell, R. M., Hertel, N. E., Wang, Z., Hutchinson, J. & Fullerton, G. D. [2006]. Calculation of effective dose from measurements of secondary neutron spectra and scattered photon dose from dynamic MLC IMRT for 6 MV, 15 MV, and 18 MV beam energies, *Medical Physics* 33: 360–368.

[13] Howell, R. M., Kry, S. F., Burgett, E., Hertel, N. E. & Followill, D. [2009]. Secondary neutron spectra from modern Varian, Siemens, and Elekta linacs with multileaf collimatorss, *Medical Physics* 36: 4027–4038.

[14] IAEA-398 [2000]. International Atomic Energy Agency, Absorbed dose determination in external beam radiotherapy: An international code of practice for dosimetry based on standards of absorbed dose to water, *Report 398* .

[15] ICRU-40 [1986]. International Commission on Radiation Units and Measurements. The Quality Factor in Radiation Protection, *Report 40* .

[16] ICRU-44 [1989]. International Commission on Radiation Units and Measurements. Tissue substitutes in radiation dosimetry and measurement, *Report 44* .

[17] IPSM-35 [1990]. Institute of Physical Sciences in Medicine, Code of practice for high-energy photon therapy dosimetry based on the NPL absorbed dose calibration service, *Physics in Medicine and Biology* 35: 1355–1360.

[18] Kase, K. R., Mao, X. S. & Nelson, W. R. [1998]. Neutron fluence and energy spectra around the Varian Clinac 2100C/2300C medical accelerator, *Health Physics* 74: 38–47.

[19] Kry, S. F., Howell, R. M., Salehpour, M. & Followill, D. [2009]. Neutron spectra and dose equivalents calculated in tissue for high-energy radiation therapy, *Medical Physics* 36: 1244–1250.

[20] Lin, J. P., Chu, T. C., Lin, S. Y. & Liu, M. T. [2001]. The measurement of photoneutrons in the vicinity of a Siemens Primus linear accelerator, *Applied Radiation and Isotopes* 55: 315–321.

[21] Lin, J. P., Liu, W. C. & Lin, C. C. [2007]. Investigation of photoneutron dose equivalent from high-energy photons in radiotherapy, *Applied Radiation and Isotopes* 65: 599–604.

[22] Ma, A., Awotwi-Pratt, J., Alghamdi, A., Alfuraih, A. & Spyrou, N. M. [2008]. Monte Carlo study photoneutron production in the Varian Clinac 2100c linac, *Journal of Radioanalytical and Nuclear Chemistry* 276: 119–123.

[23] Mao, X. S., Kase, K. R., Liu, J. C., Nelson, W. R., Kleck, J. H. & Johnsen, S. [1997]. Neutron sources in the varian clinic 2100C/2300C medical accelerator calculated by the EGS4 code, *Health Physics* 72: 524–529.

[24] Martínez, S. A., Barquero, R., Gómez Ros, J. M. & Lallena, A. M. [2011]. Neutron dose equivalent and neutron spectra in tissue for clinical linacs operating at 15, 18 and 20 MV, *Radiation Protection Dosimetry* p. doi:10.1093/rpd/ncq501.

[25] Martínez, S. A., Barquero, R., Gómez Ros, J. M. & Lallena, A. M. [2012]. Neutron dosimetry in organs of an adult human phantom using linacs with multileaf collimator in radiotherapy treatments, *Medical Physics* 39: 2854–2866.

[26] Martínez, S. A., Barquero, R., Gómez Ros, J. M., Lallena, A. M., Andrés, C. & Tortosa, R. [2010]. Evaluation of neutron production in new accelerator for radiotheraphy, *Radiation Measurements* 45: 1402–1405.

[27] McCall, R. C., Jenkins, T. M. & Shore, R. A. [1979]. Transport of accelerator produced neutrons in a concrete room, *IEE Transactions on nuclear Science* NS-26: 1593–1602.

[28] Mesbahi, A. [2006]. Development a simple point source model for Elekta SL-25 linear accelerator using MCNP4C Monte Carlo code, *Iranian Journal of Radiation Research* 4: 7–14.

[29] NCRP-79 [1984]. National Council on Radiation Protection and Measurements. Neutron contamination from medical electron accelerators, *Report 79* .

[30] Nelson, W. R., Hirayama, H. & Rogers, D. W. [1987]. The EGS4 code system, *National Accelerator Laboratory* p. SLAC Report 265.

[31] Ongaro, C., Nastasi, U. & Zanini, A. [1999]. Monte Carlo simulation of the photo-neutron production in the high-Z components of radiotherapy linear accelerators, *Monte Carlo Methods Applied* 5: 69–79.

[32] Ongaro, C., Zanini, A., Nastasi, U., Rodenas, J., Ottaviano, G. & Manfredotti, C. [2000]. Analysis of photoneutron spectra produced in medical accelerators, *Physics in Medicine and Biology* 45: L55–L61.

[33] Pelowitz, D. B. [2005]. MCNPX User's Manual Version 2.5.0, *Los Alamos National. Laboratory* pp. Report LA–UR–02–2607.

[34] Pena, J., Franco, L., Gómez, F., Iglesias, A., Pardo, J. & Pombar, M. [2005]. Monte Carlo study of Siemens PRIMUS photoneutron production, *Physics in Medicine and Biology* 50: 5921–5933.

[35] Reft, C. S., Runkel-Muller, R. & Myrianthopoulos, L. [2006]. In vivo and phantom measurements of the secondary photon and neutron doses for prostate patients undergoing 18 MV IMRT, *Medical Physics* 33: 3734–3742.

[36] S. A. Martínez Ovalle, J. M. Gómez Ros, A. M. L. R. [2011]. *Estudio Monte Carlo de la dosimetría de fotoneutrones producidos en aceleradores de uso clínico*, Documentos CIEMAT.

[37] Saeed, M. K., Moustafa, O., Yasin, O. A., Tuniz, C. & Habbani, F. I. [2009]. Doses to patients from photoneutrons emitted in a medical linear accelerator, *Radiation Protection Dosimetry* 133: 130–135.

[38] Schuhmacher, H. & Siebert, B. R. L. [1992]. Quality factors and ambient dose equivalent for neutrons based on the new ICRP recommendations, *Radiation Protection Dosimetry* 40: 85–89.

[39] Schulz, R. J., Almod, P. R., Cunningham, J. R., Garrett Holt, J., Loevinger, R., Suntharalingam, N., Wright, K. A., Nath, R. & Lempert, G. D. [1983]. American Association of Physicists in Medicine, A protocol for the determination of absorbed dose from high-energy photon and electrons beams, *Medical Physics* 10: 741–771.

[40] Vanhavere, F., Huyskens, D. & Struelens, L. [2004]. Peripheral neutron and gamma doses in radiotherapy with an 18 MV linear accelerator, *Radiation Protection Dosimetry* 110: 607–612.

[41] Zanini, A., Durisi, E., Fasolo, F., Ongaro, C., Visca, L., Nastasi, U., Burn, K. W., Scielzo, G., Adler, J. O., Annand, J. R. M. & Rosner, G. [2004a]. Monte Carlo simulation of the photoneutron field in linac radiotherapy treatments with different collimation systems, *Physics in Medicine and Biology* 49: 571–582.

[42] Zanini, A., Durisi, E., Fasolo, F., Visca, L., Ongaro, C., Nastasi, U., Burn, K. W. & Annand, J. R. M. [2004b]. Neutron spectra in a tissue equivalent phantom during photon radiotherapy treatment by linacs, *Radiation Protection Dosimetry* 110: 157–160.

Section 3

Clinical Radiotherapy

Curative Radiotherapy in Metastatic Disease: How to Develop the Role of Radiotherapy from Local to Metastases

Chul-Seung Kay and Young-Nam Kang

Additional information is available at the end of the chapter

http://dx.doi.org/10.5772/56556

1. Introduction

Metastasis is the leading cause of cancer death and in patients with proven distant metastases from solid tumors, it has been a notion that the condition is incurable and the treatment is usually conducted with palliative intent, with rare exceptions. Treatment predominantly involves the use of systemic chemotherapy, targeted radiotherapy or local measures typically reserved for symptom relief (Argiris, 2004; Escudier, 2007; Hurwits, 2004). Chemotherapy is delivered without expectation of long term survival, except for highly chemosensitive malignancies, such as leukemia, lymphoma, and germ cell tumors. According to the conventional treatment strategy for solid tumors, the presence of metastatic disease is a contraindication for local therapy because it is believed that these tumor cells have already spread systemically. However, from the viewpoint of reducing tumor burden, local therapy may be an adequate strategy when the target lesions account for the major portion of the total tumor volume. The local therapies are metastasectomy, heating or cooling, and radiotherapy.

In a subset of patients with a limited number of metastases or oligometastases, local ablative therapy, such as surgical resection, can potentially yields favorable outcomes. For instance, in surgical literature, it has been demonstrated that surgical resection of limited lung and liver metastases has results in prolonged survival and possibly cure in a significant proportion of patients with oligometastases (Fong, 1999; Friedel, 1994). International registry of lung metastases reported that lung metastasectomy is a safe and curative procedure in selected patients with disease free interval (DFI) ≥ 3 years, single lesion and germ cell tumor (Pastorino, 1997). And the lung and liver metastasectomy is a surgical approach used in colon and breast cancers, with upto 22% of colon cancer patients surviving 10 years and 35-46% of breast cancer patients surviving upto 5 years (Fong et al, 1999; Friedel et al, 1994; Pocard et al, 2001).

However, resection may not be feasible in patients of extremely advanced age or with poor cardiopulmonary function or multiple comorbidities because of the risk of significant morbidity and mortality in these settings. In such patients, external radiotherapy is often the treatment of choice. However, this treatment also has its drawbacks, including the potential to damage adjacent or nearby structures and its association with a local failure rate that is higher than that was seen in resection. To minimize collateral injury to normal tissues, adequate fractionation (e.g. 1.8-2.0 Gy/fraction) over 6 to 7 weeks is commonly used. The use of radioablative treatment such as using stereotactic radiosurgery (SRS) or stereotactic body radiotherapy (SBRT) to overcome problems with normal tissue injury in patients with medically inoperable metastatic tumor has now been actively studied at many institutions.

2. Oliogmetastases

2.1. Definition

In 1889, Paget's "seed and soil" hypothesis stated that the development of a metastasis depends on cross talk between selected cancer cells (the seed) and a specific organ microenvironment (the soil) (Paget, 1889). This means that successfully establishing a distant metastasis depends on certain properties of the host organ as well as those of tumor cells. Dissemination of tumor cells in general circulation does not necessarily mean that wide spread metastatic disease will always develop. This hypothesis is still widely accepted. Observing the natural history of breast cancer, Hellmann and Weichselbaum hypothesized the existence of an intermediate state between widespread metastatic disease and locally confined disease and coined the term "oligometastases" (Hellman & Weichselbaum, 1995). Thus, local control of oligometastatic disease may allow better systemic control. In addition, thanks to the evolution of radiologic imaging technique, detection of metastasis at a size previously impossible to be detected may result in under treatment and an effective chemotherapy may downstage these metastatic diseases to oligometastases.

2.2. Clinical significance

Clinically, there are two types of oligometastases. The De novo type is the tumor early in the evolution of metastatic progression producing metastasis that are limited in number and location, and the induced type is generated when effective systemic chemotherapy eradicates the majority of metastatic deposits in a patients with wide spread metastatic disease (MacDermed, 2008).

In a retrospective study, Mehta et al tracked the number of individual metastatic sites and the number of organs involved using serial computerized tomography of the body in 38 patients with stage IIIb or IV non small cell lung cancer treated with chemotherapy. Seventy four percent of patients (n=28) had a metastatic disease limited to 1-2 organs and 50% (n=29) had a disease limited to the primary tumor and three or less metastatic lesions at presentation. Fifty percent (n=19) had stable (n=12) or progressive (n=7) disease in initially involved site without development of new metastatic lesion. Among the 17 patients with four or fewer metastatic sites with no pleural effusions, 65% (n=11) had stable or progressive disease in initially involved sites without development of new metastatic lesions (Mehta, 2004). The results of this study suggest that a subset of patients with oligometastases from lung cancer may benefit from a combination of systemic chemotherapy and local aggressive therapy. In another study, records of 387 patients with advanced non small cell lung cancer were reviewed and 64 patients with measurable advanced stage non small cell lung cancer who received first line systemic chemotherapy and follow up were identified. Thirty four patients were deemed theoretically SBRT eligible. Disease in the lung and liver was limited to ≤ 3 sites each. Among the 34 SBRT eligible patients, the pattern of failure were local only in 68%, distant only in 14%, and mixed in 18%. The time to first progression was 3.0 months in those with local only failure (Rusthoven, 2009). The results of this study suggest that SBRT may improve the time to first progression in a significant proportion of patients with metastatic non small cell lung cancer. After all, because any patient with oligometastatses may exist in a spectrum between orderly metastatic progressions and wide spread occult disease, the role of the local modality to ablate oligometastases need to be determined.

3. Stereotactic Body Radiation Therapy (SBRT)

3.1. Definition and characteristics

The scientific study and clinical practice of oncology have progressed remarkably in recent years. Insights into molecular interactions occurring within a cancer cell have been translated into novel medical treatments, and a variety of technological advances have allowed new surgical and radiotherapeutic techniques. Within the discipline of radiation oncology in particular, the fusion of state-of the-art tumor imaging with precision radiation treatment delivery systems has created an opportunity to shift from the classic radiation therapy paradigm of administering thirty or more individual allow-dose treatments toward briefer, more intense, and more potent regimens in which a much higher dose per treatment is used for greater clinical effect. Stereotactic body radiation therapy (SBRT) refers to an emerging radiotherapy procedure that is highly effective in controlling early stage primary and oligometastatic cancers at locations throughout the abdominopelvic and thoracic cavities, and at spinal and paraspinal sites. The major feature that separates SBRT from conventional radiation treatment is the delivery of large doses in a few fractions, which results in a high biological effective dose (BED). In order to minimize the normal tissue toxicity, conformation of high doses to the target and rapid fall-off doses away from the target is critical. The practice of SBRT therefore requires a high level of confidence in the accuracy of the entire treatment delivery process.

In SBRT, radiation is targeted almost exclusively to the tumor, while tissues not grossly involved with the tumor are spared. However, unique radiobiology of SBRT that ensures maximal tumor control but minimal normal tissue complication is what really sets SBRT apart from other radiotherapy techniques. Additional defining characteristics of SBRT include the abilities to securely immobilize the patient for the typically long treatment sessions; to accurately duplicate patient position between simulation and treatment; to minimize normal tissue exposure through the use of multiple- or large angle, arcing, small-aperture fields; to rigoursly account for organ motion; to stereoctactically register tumor target and normal tissue structures; and to deliver ablative dose fractions with subcentimeter accuracy to the patient (Timmerman, 2007).

Immobilization and repositioning devices include the Elekta Stereotactic Body Frame™ (Elekta, Norcross, Ga., USA), the Leibinger stereotactic body fixation system (Stryker, Kalamazoo, Mich., USA), and the Medical Intelligence Bodyfix™ system (Medical Intelligence, Schwabmuenchen, Germany).

Several systems provide one or another solution to the problem of respiratory motion. A breath-hold device is the Active Breathing Coordinator™ (Elekta), which allows coordination of beam-on time during a fixed level of inspiration; a respiratory gating system is the RPMTM (Varian, Palo Alto, USA) which tracks inspiration and expiration and turns the accelerator off when indicators predict that the tumor position is outside of an acceptable range of distance from baseline; a another gating system is the ANZAI (Anzai, Japan).

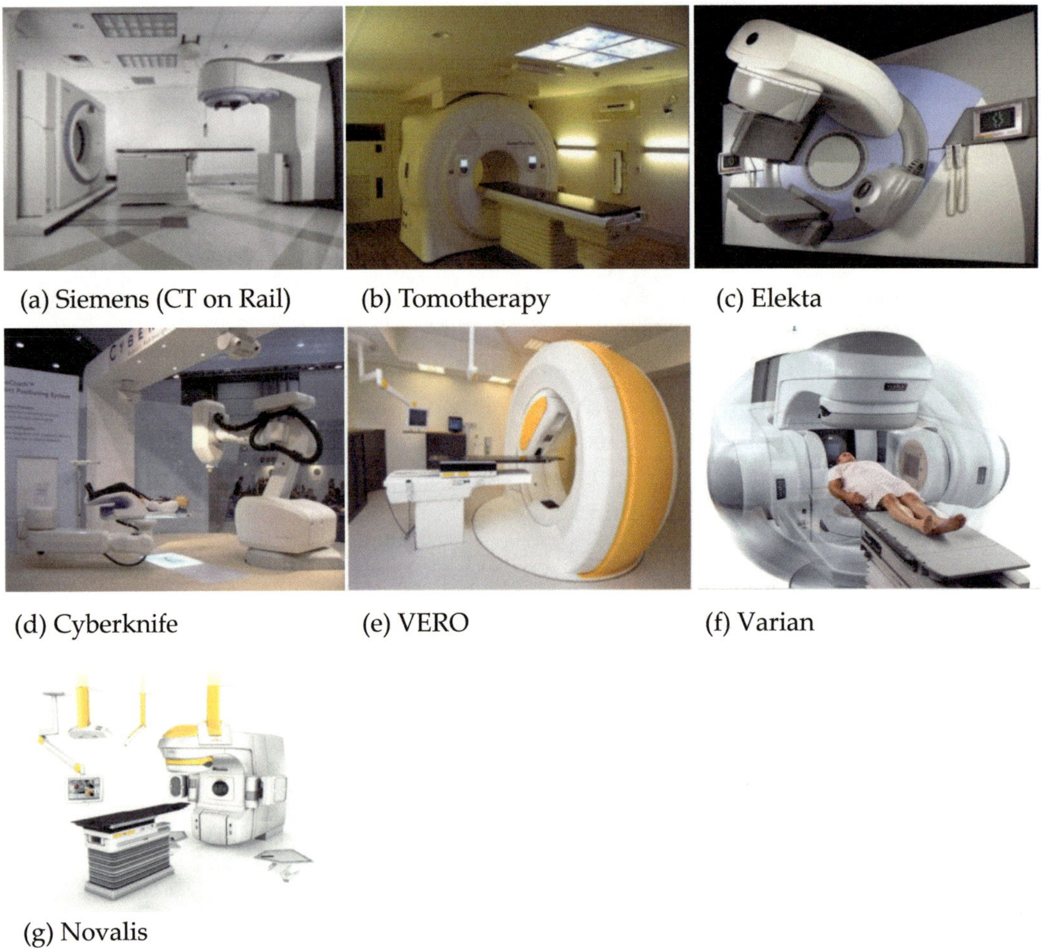

(a) Siemens (CT on Rail) (b) Tomotherapy (c) Elekta

(d) Cyberknife (e) VERO (f) Varian

(g) Novalis

Figure 1. Various system of Stereotactic Body Radiation Therapy (SBRT)

Also now available for purchase are specialized SBRT-ready linear accelerators that combine capacity for image-guided radiotherapy with compatibility with modern immobilization and respiratory motion solution technology. The Novalis™ (BrainLAB, Inc.), Elekta Synergy™, Varian Trilogy™, Siemens Atiste™ System, Tomotherapy HiArt™ System (TomoTherapy, Madison, Wisc., USA), and Cyberknife™ (Accuray, Sunnyvale, Calif., USA) are linear accelerators for SBRT (Fig. 1).

3.2. Radiobiologic aspect of Hypofractionated SBRT

The most prevalent method of radiotherapy in the past 100 years of radiation oncology has been a strategy known as protracted fractionation in which daily small doses of radiation (e.g. 1.8~4 Gy) are delivered repeatedly over many days. The basis of this method of radiotherapy was that normal tissue repairs sublethal injury between fractions better than does tumor tissue. With the advent of SRS to treat intracranial tumors, an alternatively strategy of giving an ablative dose (e.g. 12~30 Gy) was born. A SBRT is an extension of this technique to deliver ablative radiotherapy (8~30 Gy) to extracranial sites. When alternate fractionation schemes are considered, we need some model for calculating isoeffect doses and a linear quadratic (LQ) formalism is most commonly used for quantitative prediction of dose/fractionation dependencies. The LQ model approximates clonogenic survival data with a truncated power series (second order polynomial) expansion of natural log of S (surviving proportion) as follows (see Equation 1).

$$\ln S = -\alpha * d - \beta * d^2 \qquad (1)$$

The d is daily dose and α & β are expansion parameters: α is the slope of the survival curve at the limit $d \rightarrow 0$, and β is the parameter determining the relative contribution from the quadratic component (Fig. 2A). This model was initially derived to fit experimental observation of the effects of dose and fractionation on cell survival, chromosomal damage, and acute radiation effects. Later, some ascribed underlying biological mechanism to the mathematical terms, primarily the formation of single- and double-strand break in DNA. The LQ model has been useful for predicting and understanding the effects of conventional fractionated radiotherapy. The biological effective dose (BED) is a characteristic dose value that facilitated comparisons between the effects of different dose fractionation schemes. The BED is defined as the total dose delivered in an infinite number of infinitesimally small dose fractions that has the same biologic effect as the dose fractionation scheme in question and described as $BED=D*[1+d/(\alpha/\beta)]$.

This BED based on LQ model is known reasonably predictive of dose response relation, both in vivo and vitro, in the dose per fractions range of 2 to 15 Gy (Brenner, 2008), however, the LQ model predicts a continuously bending curve in the high dose range and experimentally measured data have decidedly shown a linear relationship between the dose and log of the proportion of surviving clonogen. In addition, in the early phase of its development, one of the developers of the LQ model stated that "LQ is not intended for doses higher than 8-10 Gy. In any case, LQ is simply a loose dose approximation to equation

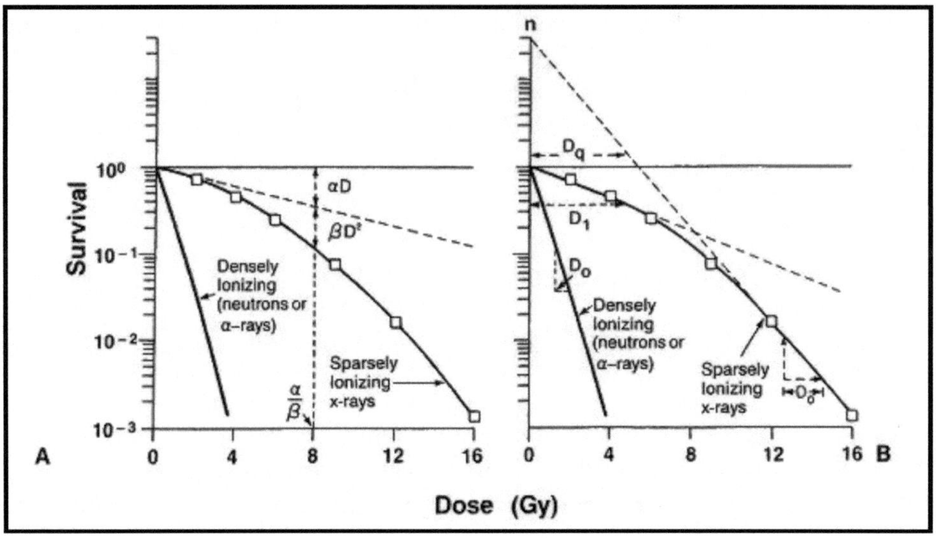

Figure 2. Shape of survival curve for mammalian cells exposed to radiation. The fraction of cells surviving is plotted on a logarithmic scale against on a linear scale. For α-particles or low energy neutrons (said to be densly ionizing), the dose-response curve is a straight line from the origin (i.e., survival is an exponential function of dose). The survival curve can be described by just one parameter, the slope. For X- or γ-rays (said to be sparsely ionizing), the dose-response curve has an initial slope, followed by a shoulder; at higher dose, the curve tends to become straight again. A: The linear quadratic model. The experimental data are fitted to a linear quadratic function. There are two components of cell killing: One is proportional to dose (αD); the other is proportional to the square of the dose (βD^2). The dose at which the linear and quadratic components are equal is the ratio α/β. The linear quadratic curve bends continuously but is a good fit to experimental data for the first few decades of survival. B: The multitarget model. The curve is described by the initial slope (D_1), the final slope (D_0), and a parameter that represents the width of the shoulder, either n or D_q (Hall, 2006).

that do become straight exponential at higher dose" (Hall, 1993). Thus, LQ model overestimates the effect of radiation on clonogenicity in the high dose commonly used in SBRT and inappropriate to apply at the high doses per fraction encountered in radiosurgery because it (1) does not accurately explain the observed (in vivo) data; (2) was derived largely from, in vitro, rather than in vivo, observations and, thus, does not consider the impact of ionizing radiation on the supporting tissues; (3) does not consider the impact of subpopulation of radioresistant clonogens (ie, the "cancer stem cell" response); and (4) creates a "false belief" that this simplified model represent an absolute truth (Kirkpatrick, 2008). Substantial modifications are needed to apply the LQ model to the SBRT regimen; at which point the model loses its simplicity and natural appeal (Guerrero & Li, 2004). The multitarget model (Fig. 2B) provides an alternative description of the clonogenic survival as a function of radiation dose and is still valuable because it fits the empirical data well, especially in the high dose range. In a study of University of Texas, Park et al proposed a new model, universal survival curve (USC), to reconcile the strengths of these LQ model and multitarget model into single, unifying model and stressed that the proposed survival curve model (Fig. 3) (Park, 2008).

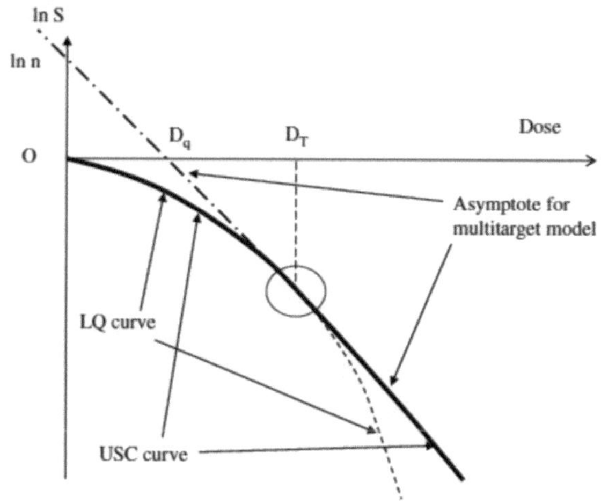

Figure 3. Universal survival curve (USC) and transition between dose range at which linear quadratic (LQ) model is valid and dose range at which multitarget model is valid. Below transition dose D_T, USC curve is identical with LQ model curve and above D_T, USC curve is identical with terminal linear portion of multitarget model curve.

The USC model can be used to derive isoeffective relations (equivalent dose function) of any arbitrarily fractionated RT. For SBRT, a novel concept of the single fraction equivalent dose (SFED) can serve as an alternative and more intuitive way to compare different dose fractionation schemes. SFED was defiend as the dose delivered in a single fraction that would have the same biologic effects as the dose fractionation scheme in question. For total dose D given in n fractions, each fraction with the dose, d, SFED is determined by the intersection line crossing the effective survival curve at D=d*n (Fig. 4) (Park, 2008).

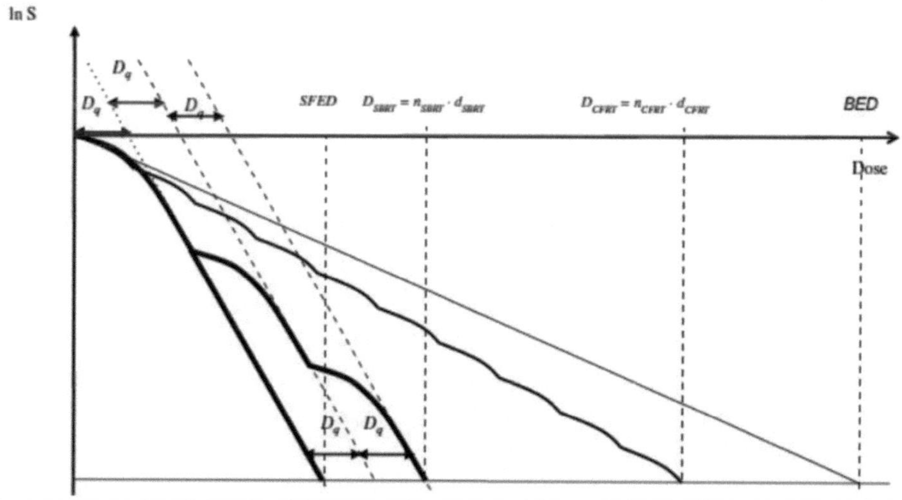

Figure 4. Graphic representation of stereotactic body radiotherapy (SBRT) and conventionally fractionated radiotherapy (CFRT) effective survival curves compared with single fraction equivalent dose (SFED) and biological effective dose (BED). Note, D_{CFRT} and D_{SBRT} are always situated between biologically equivalent dose parameters of SFED and BED.

The definition of BED and SFED using the USC curve remain applicable for any fractionation regimen. The ln S in fractionated conventional fractionated radiation therapy (CFRT) and SBRT is calculated (see Equation 2A and 2B).

$$\begin{aligned} \ln S &= -n*\left(\alpha * d + \beta * d^2\right) \text{ if } d \le D_T && \text{(A)} \\ &= -n*\left(1/D_0 * d - D_q/D_0\right) \text{ if } d \ge D_T && \text{(B)} \end{aligned} \qquad (2)$$

Thus, Equations. 2A and 2B are arranged to reflect the common clinical practice in designing dose fraction scheme in which d_{CFRT} and n_{CFRT} is varied for CFRT, and n_{SBRT} is fixed and d_{SBRT} is varied for SBRT. By letting d_{CFRT}=2 Gy, the equation for D_{CFRT} can be used to calculated the standard effective dose (SED), total dose in 2 Gy fractions with equivalent effect (see Equation 3A and 3B).

$$\begin{aligned} D_{SBRT} &= \alpha * D_0 * D_{CFRT} * \left[1 + d_{CFRT}/(\alpha/\beta)\right] + n_{SBRT} * D_q && \text{(A)} \\ D_{CFRT} &= 1/\alpha * D_0 \left\{\left[D_{SBRT} - \left(n_{SBRT} * D_q\right)\right] / \left[1 + d_{CFRT}/(\alpha/\beta)\right]\right\} && \text{(B)} \end{aligned} \qquad (3)$$

From the report of 12 non small cell lung cancer cell lines from National Cancer Institute, the value for α, D_0 or D_q was obtained by determining the arithmetic mean values of each parameter. The range of D_q was wide, between 1.5 Gy and 2.5 Gy. The mean value for α, D_0, and D_q was 0.33 Gy^{-1}, 1.25 Gy and 1.8 Gy, respectively (Carmichael, 1989; Morstyn, 1984). The transition dose, D_T was calculated to be 6.2 Gy, reassuring results, because the dose per fraction used in CFRT is < 6.2 Gy and SBRT regimens commonly use a dose per fraction > 6.2 Gy. This USC is a new model that may offer a superior description of the mammalian cell survival curve in the ablative dose range beyond the shoulder, on the same time, preserving all the strengths of the LQ model in the low dose range (around the shoulder). However, the true survival of in vivo tumors depends on multiple factors that cannot possibly be contained in simplified mathematical formulas. The only way to truly know the tumor control rates or the tolerance of different fractionation schemes is through performing prospectively designed trial.

3.3. Variable imaging technique to guarantee high accuracy radiotherapy

In delivering SBRT, many commercially available units can be utilized. Sophisticated image guidance is a common feature to these treatment units. Units equipped with online image-guided radiation therapy (IGRT) capability minimize the uncertainty associated with tumor localization. In-house developed systems such as RT-RT and CT-on-rails were employed prior to the widespread availability of in-treatment-room imaging. Recent developments have spread the availability of in-treatment-room imaging to many facilities.

3.3.1. Cone beam CT Linear accelerator IGRT

The first commercially available cone beam CT (CBCT) IGRT system was the Elekta Synergy™ (Elekta, Crawley, UK), the other medical linear accelerator (linac) manufacturers have also now embraced the IGRT concept and have either produced their own version of an IGRT linac, Varian Trilogy™ (Varian Medical Systems, Palo Alto, Calif., USA), or are in the process of such developments, Siemens ARTÍSTE™ (Siemens Medical Solutions USA, Inc., Malvern, Pa.,USA). The Synergy and Trilogy consists of a retractable kV X-ray source, an amorphous silicon flat panel imager mounted on the linear accelerator perpendicular to the radiation beam direction, and a software module (referred to as the XVI system). The system provides planar, motion, and volumetric images.

For CBCT image acquisitions, the gantry is rotated around the patient for a preset angle (between 180° and 360° to allow sufficient data acquisition) and images are acquired via an amorphous silicon panel. Volumetric image reconstruction is performed simultaneously with the acquisition to expedite the process. The reconstructed three-dimensional geometry is subsequently registered with the reference geometry planning image, either manually or automatically (using either soft tissue or bone mode). For some disease sites, such as prostate cancer, the soft tissue mode is conceptually ideally suited, since the prostate often moves relative to the bones. However, at present, it is difficult to visualize the prostate in all cases, and thus implanted radiopaque seeds are used to make the registration process more efficient. Based on the registration, the difference between the data sets is calculated and displayed as translation along and rotation about the three axes. Subsequent treatment table adjustments are made and the patient treated. One can clearly appreciate that CBCT-based IGRT shows great potential for objective, precise positioning of patients for treatment, matching the treatment setup image model to that of the planning image model. It remains to be determined exactly which imaging features on the integrated CBCT linacs (i.e., kVp CBCT, planar, motion, and MV electronic portal imaging device) are best suited for a particular disease site.

3.3.2. Helical tomotherapy IGRT

Helical tomotherapy was first proposed by Mackie et al. in 1993 and is now commercially available as the TomoTherapy HI -ART system (TomoTherapy, Inc., Madison, Wisc., USA). A short in-line 6-MV linac (Siemens Oncology Systems, Concord, Calif., USA) rotates on a ring gantry at a source-axis distance of 85 cm. The IMRT treatment is delivered while the patient support couch is translated in the y-direction (toward the gantry) through the gantry bore, in the same way as a helical CT study is conducted. In the patient's reference frame, the treatment beam is angled inwards along a helix with the midpoint of fan beam passing through the center of the bore. Similar to helical CT, the treatment beam pitch is defined as the distance traveled by the couch per gantry rotation, divided by the field width in the y-direction. The width of the beam in the y-direction is defined by a pair of jaws that is fixed, for any particular patient treatment, to one of three selectable values (1, 2.5 or 5 cm). Laterally, the treatment beam is modulated by a 64 leaf binary multileaf collimator, whose leaves transition rapidly between open and closed states providing a maximum possible open lateral field length of 40 cm at the bore center. Highly modulated treatments can achieve great conformality, though they inevitably take longer to deliver. A helical MV CT image is acquired prior to treatment each day using the on-board xenon CT detector system and the 6-MV linac (detuned to 3.6 MV). Registration software is provided to compare the daily patient setup image with the stored prescription CT planning image. After image registration, table adjustments are then automatically made and the patient is then treated.

3.3.3. Megavoltage Cone Beam CT Linear Accelerator IGRT

The only MV Cone Beam CT (CBCT) system currently available is the most recent addition to the family of in-room 3D systems designed for IGRT. The MV CBCT imaging system consists of a 6-MV x-ray beam produced by a conventional linear accelerator (Oncor, Siemens AG, Erlangen, Germany) equipped with an amorphous-silicon EPID (AG9-ES, PerkinElmer Optoelectronics, Waltham, MA., USA) flat panel detector. The system is controlled by a computer workstation (Syngo Coherence RTT, Siemens AG, Erlangen, Germany) that is responsible for all tasks related to portal or MV CBCT imaging, including calibration of the system, quality assurance, image acquisition, and image registration (2D or 3D) for patient alignment. The use of MV photons for imaging is a departure from conventional preferences of using kilovoltage (kV) photons, which have resulted in superior image quality for diagnostic purposes. The MV CBCT system is capable of measuring setup errors of fiducials in an anthropomorphic head phantom with submillimeter accuracy and reproducibility. The gantry rotates in a continuous 200° arc (270° to 110°) while acquiring one low-dose portal image per degree. The 200 projection images acquired are then used for MV CBCT reconstruction, which is completed approximately 2 minutes after the starts with an automatic registration, based on a maximization of mutual information algorithm, which utilizes all information in both 3D images to maximize the alignment of similar structures.

Routine quality assurance on the system has also demonstrated that the calibrated MV CBCT imaging isocenter remained within 1 mm to the machine treatment isocenter over a period of 1 year. As for the field-of-view, anatomical information situated in a 27 X 27 X 27 cm^3 volume centered at isocenter is reconstructed in the MV CBCT system with a half-beam acquisition mode should increase the reconstruction size in the axial plane by up to 40 cm.

3.3.4. Vero IGRT

Vero SBRT is specifically designed for IGRT and a new type of 6 MV linac with attached MLC is mounted on an O-ring gantry. The MLC consists of sixty 5-mm-leaves and produces a maximum field size of 150 x 150 mm^2. The gantry rotates 360 degree and the horizontal axis, similar to a C-arm linac platform, but additionally allows rotation about the vertical axis. The system incorporates the MHI-TM2000 linear accelerator and sophisticated software to deliver radiation therapy. The system is equipped with a dual orthogonal kV imaging systems attached to the O-ring at 45 from the MV beam. This imaging system allows simultaneous acquisition of orthogonal X-rays images and fluoroscopy. Also kV CBCT imaging is available. Vero SBRT dynamically contours the treatment beam exactly to the tumor from every angle as the machine moves around the patient. Furthermore, Vero's technology allows clinicians to dynamically treat with a moving beam in order to spare surrounding healthy tissue and organs while maximizing such as x-ray, CT and fluoroscopy, so that clinicians can modify their plans during treatment as needed. The targeted beam adapts to breathing and other body movements to maintain safe, complete and accurate dose delivery.

3.3.5. Electromagnetic tracking

One of the earliest applications of electromagnetic tracking in RT was for the nonradiographic localication of interstitial abdominal implants for intraoperative high-dose-rate (HDR). In this application a then-commercially available 3SPACE-FASTRAK system (Polhemus Inc, Colchester, VT., USA) was configured to fit in the lumen of a catheter. The system was then used to measure the spatial path of all catheters by inserting the wired sensor sequentially into each catheter. This information was then used by the planning system to accurately determine and calculate dwell positions and times. The stated accuracy of the system was a root mean square (RMS) of 0.8 mm, but measurements in the operating environment found the RMS accuracy to be 0.38 mm in the absence of metallic surgical retractors and 0.70 mm in the presence of three retractors, with maximum absolute errors of 2.1 mm or less.

In 2000, the Paul Scherrer Institute reported on an electromagnetic tracking system they had developed for real-time (50 Hz) target volume tracking during proton therapy with continuous spot scanning delivery. This system consisted of an external magnetic field generator, a wired implantable sensor, and the associated signal processing electronics. When compared with an optical tracking device with 30 μm accuracy, the RMS spatial accuracy was reported to be 1 mm to 2 mm, whereas the RMS angular accuracy of determining the orientation of the dipole was 0.5 to 1 degree. The system's ability to track and gate was tested in a moving phantom and qualitatively shown to very nearly restore the dose distribution to the planned static distribution when a 3-mm gating window was implemented. The technology for this system was developed by a spin-off company from the Paul Scherrer Institute called Mednetix AG, which was acquired by Northern Digital Inc (Waterloo, ON, Canada). Further development efforts have focused on a wired electromagnetic tracking system for guidance of medical instruments, which is commercially available in the Aurora system.

3.3.6. Cyberknife IGRT

The use of a small X-band linear accelerator mounted on an industrial robot was first developed for radiosurgery. The robot provides the capability of aiming beamlets with any orientation relative to the target volume. The system uses two ceiling-mounted diagnostic X-ray sources, and amorphous silicon image detectors mounted flush to the floor. The treatment is specified by the trajectory of the robot and by the number of monitor units delivered at each robotic orientation. During the patient's treatment, the Cyberknife System correlates live radiographic images with preoperative CT or MRI scans in real time to determine patient and tumor position repeatedly over the course of treatment. More details are provided by users of this system in subsequent articles in this volume.

3.4. Compensation of respiratory motion of the tumor and internal organ

SBRT requires precise delineation of patient anatomy, targets for planning, and clear visualization for localization during treatment delivery. Three-dimensional data sets assembled from CT or 4DCT for visualizations and dose calculation and/or MRI and positron emission tomography (PET) images assist in target and visualization for SBRT.

3.4.1. Four-Dimensional CT scanning

Respiratory correlated 4DCT was developed over the past several years to address the issues of respiratory motion in radiotherapy targeting (Rietzel, 2005). Respiration-correlated CT uses a surrogate signal, such as the abdominal surface, respiratory air flow, or internal anatomy to provide a signal that permits resorting of the reconstructed image data, resulting in multiple coherent spatiotemporal data sets at different respiratory phases. The scan time for 4DCT with multi-slice scanners is on the order of a few minutes, and post-processing takes an additional 30 min if manual phase selection is required. The output of this process is typically 10 CT volumes, each with a temporal resolution of approx. 1/10 of the respiratory period. 4DCT uses multi-slice CT scanners combined with a respiratory surrogate to develop a series of 3DCT scans each representing the patient in a different respiratory phase. The entire 4DCT dataset can be used to determine an envelope of tumor motion which can be expanded to include areas of subclinical disease resulting in an internal target volume (ITV) (ICRU 1999) which can be used as the treatment target. Alternatively, select phases from the 4DCT can be used to determine an ITV that only covers a select range of respiratory phases (i.e. 40%-60% corresponding to a ± 10% window around end exhalation) that would be the target for gated treatments. The most common form of motion management used in RTOG studies to date and also at many experienced centers using SBRT across the world has been chest wall breathing with abdominal compression. Chest wall breathing exerts forces on the intrathoracic tissues in multiple opposing directions in contrast to the mostly craniocaudal force vectors associated with diaphragmatic breathing. As a result, the amplitude of tumor motion with chest wall breathing can be significantly decreased. With this technique, the patient is first coached to expand the lungs using their upper chest wall rather than by moving their diaphragm toward their abdomen.

The 4DCT implementation relies on sensing the respiratory phase by using the Varian RPM system or Anzai system. 4DCT provides an imaging tool to quantify and characterize tumor and normal tissue shape and motion as a function of time. This provides the radiation oncologist and treatment planner with information essential in the design of an aperture that more adequately covers the internal target volume (assuming respiration during treatment is reproducible to that during CT simulation). 4DCT data can also be used as input in making treatment decisions on when to intervene with gating or other motion management strategies. In addition, the 4DCT data can be used as direct input into four-dimensional treatment planning, and to generate time-varying dose-volume histograms or isodose distributions. An effective method of conveying the utility of 4DCT is through computer animation. Dr. Choi et al. [pers. commun., 2005] have found that approximately one half of patients with early-stage disease have motion of less than 10 mm during quiet breathing (in approx. 100 cases). Seppen woolde et al. reported on the motion of 21 lesions in 20 patients and found a mean motion of 5.5 mm in the craniocaudal direction (data ranged from 0 to 2.5 cm). Average periodicity was observed to be 3.5 s, and ranged from 2.8 to over 6 s. The clinical importance of 4DCT is that it provides insight into patient-specific organ motion.

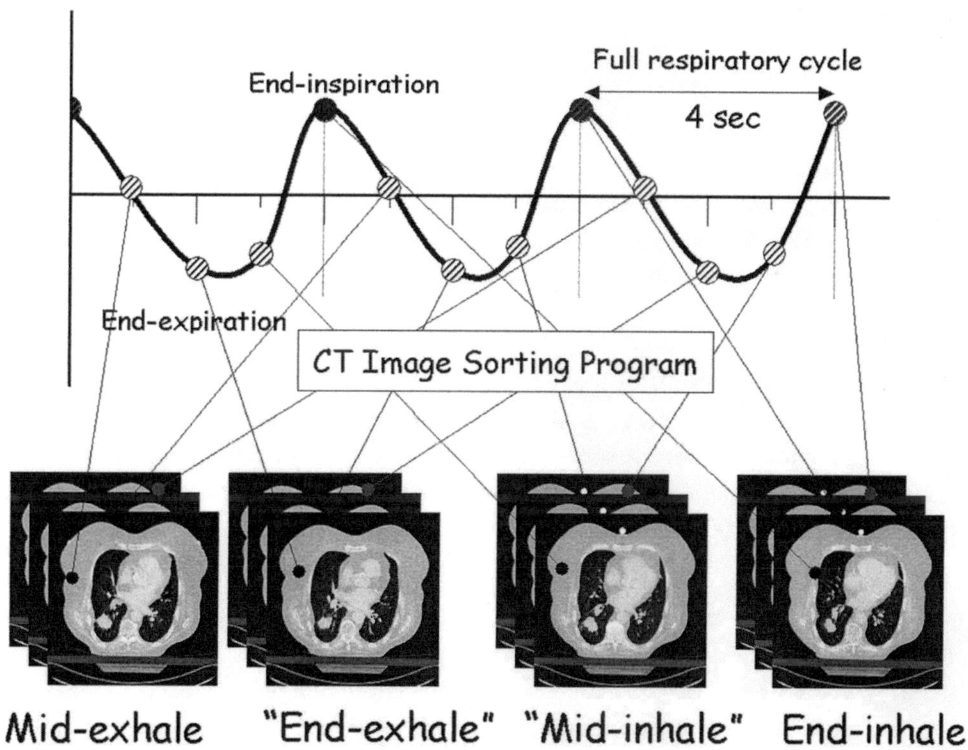

Figure 5. Overview of respiratory phase "bin" generation from four-dimensional computed tomography data.

3.4.2. Image registration & Motion management

SBRT treatments do not use invasive external frames. A body frame system has been developed that incorporates several features to ensure reproducible setup, including a vacuum bag that is fit to the patient at the time of simulation, a scale that facilitates reproducible positioning of the patient in the frame, an abdominal compression paddle to restrict abdominal motion, and external fiducial markers to improve setup accuracy (Lohr, 1999; Hadinger, 2002). This system is particularly useful when the patient is to be imaged in one room and the entire patient/body frame system is moved to the treatment room. Without a body frame, either implanted fiducial markers or in-room volumetric imaging is required for accurate internal soft tissue-based setup.

One technique for minimizing the effects of respiratory motion is to activate the radiation beam only when the tumor is at a predetermined location in the respiratory cycle. This is referred to as respiratory gating (Shirato, 2000; Starkschall, 2004; Nelson, 2005; Underberg, 2005). The use of gating requires some measure of the tumor location within the respiratory cycle, which can be done directly but is more often done through some respiratory surrogate such as abdominal height or diameter. Spirometry has also been used to gate based on tidal volume (Zhang, 2003).

Alternative motion management techniques include dynamic gating and breath-hold techniques. During dynamic gating the patient is allowed to breath normally with or without audio or visual coaching and the radiation beam is activated only when the patient reaches the planned points in their respiratory cycle. Breath-hold gating requires the patient to hold their breath at a given abdominal height or tidal volume generally with the aid of visual feedback and the radiation beam is activated only when the patient is holding their breath in this target position. The breath-hold can either be voluntary or assisted with an occlusion valve. Breath-hold has several benefits over dynamic gating including the ability to do volumetric imaging over a series of breath holds, longer irradiation times to allow radiotherapy beams to stabilize, and the ability to expand the lungs and give more fall-off distance between the target and nearby critical structures.

Gating is performed with real time or near time verification of the target position in the gate with in-treatment-room imaging. An early example of in-treatment-room imaging was developed by Shirato et al. (Shirato, 1999) who developed a real-time tumor tracking method in which four sets of x-ray tubes and fluoroscopic imagers are used to measure the position of four implanted radiopaque markers relative to the isocenter. The linear accelerator was configured so that it irradiated the tumor only when the markers were within certain coordinates. This system is effective for the treatment of lung tumors but requires the invasive implantation of fiducial markers. In addition this system has not become commercially available. A similar method is used by two commercially available stereotactic systems, Novalis®/Exactrac® (BrainLAB, Feldkirchen, Germany) (Yin, 2002) and Cyberknife® (Accuray, Sunnyvale, CA) (Adler, 1997). These systems both have room mounted orthogonal x-ray systems that can observe the patient's anatomy in the treatment position. Implanted fiducial markers are required for all lung tumors on the Novalis® system but the Cyberknife® can use either fiducial markers or direct imaging depending on the tumor location. The Novalis® system does not employ real-time tumor tracking but rather relies on a relationship between external surrogates and the tumor position developed immediately prior to treatment. Cyberknife® can either confirm the position of the target at regular intervals during treatment or utilize a respiratory tracking system that continuously synchronizes beam delivery to the motion of the target.

Non-radiographic localization was investigated by Balter et al. (Balter, 2005) who studied the use of the Calypso™ 4D system for patient positioning based on real-time localization of implanted electromagnetic transponders (beacons). This study demonstrated the accuracy of the system before clinical trials were conducted. The system consists of 5 major components: wireless transponders, a console, a detector array, a tracking station and infrared cameras. The array emits electromagnetic radiation that excites the implanted transponders. Due to the resonance response the array can locate the 3D coordinates of the wireless transponders. The infrared cameras allow the registration of the position of the array relative to the isocenter of the linear accelerator. This system offers the potential for real-time tracking and is commercially available for prostate but not yet for other body sites including the lungs.

3.5. Quality Assurance of SBRT

An important goal of a quality assurance (QA) program is to instill confidence that patients are receiving their prescribed treatments accurately. The goal should not be simply getting through some mandatory tests as quickly and painlessly as possible. Unfortunately, many catastrophic events are produced by failures happening at a moment that cannot be predicted or caught by routine quality control (QC) procedures. As there are built-in interlocks in treatment devices, most failures occur in human processes rather than in equipment. Finding the proper balance between effort spent on specific QC procedures and effort spent on an overall quality management program is major challenge at most institutions.

The QA program for an SBRT process based on IGRT system must evaluate the entire treatment process, including patient immobilization, setup, simulation imaging, treatment planning (including the production of reference images to guide corrections), verification imaging, image registration, patient position correction, and treatment. Tests that assess the entire process from beginning to end inspire confidence that the overall process is accurate and robust. In such as study, planning images of a phantom are acquired and transferred to the treatment planning system. A treatment plan is designed and reference images are produced. The phantom is then taken to the treatment unit and positioned for treatment. A verification images is acquired and registered to the reference image. Any necessary corrections to the position are made. Treatment is then delivered and measured by using an ion chamber and/or film. The dosimeter readings are compared with the expected values from the treatment plan. The frequency of such tests should be based upon an analysis of system stability during the initial operation of the SBRT system. Also, the image-guidance procedures should be reviewed on a regular basis to ensure that the procedures are consistent with the initial design and to initiate appropriate changes if necessary. The review results and any new changes must be communicated among staff. The review of the guidance elements of a patient's treatment can be integrated into the institutional chart rounds and quality control programs to verify that the image guidance procedure is operating correctly.

Conventional linear accelerator-based IGRT consists of imaging in the treatment room during a course of radiotherapy. Planar (two-dimensional [2D]) and volumetric (three-dimensional [3D]) imaging are used for repositioning the patient immediately prior to treatment. The common elements of a QA program include: (1) safety and functionality, (2) geometric accuracy (agreement of MV and kV beam isocenters), (3) image quality, (4) registration and correction accuracy, and (5) dose to patient and doseimetric stability.

Helical Tomotherapy requires synchrony of gantry rotation, couch translation, linear accelerator pulsing, and opening and closing of the binary MLC leaves used to modulate the radiation beam. The accuracy of this highly dynamic treatment process depends on the correct performance of the radiation source, MLC, gantry, and couch table. The dose delivered to the patient depends on the static beam dosimetry, system geometry, system dynamics, and system synchrony. Systematic QA of the system dynamics and synchrony has been suggested, which includes jaw width constancy, actual fraction of time leaves are open, couch drive distance, speed and uniformity, linear accelerator pulsing and gantry synchrony, leaf opening and gantry synchrony, and couch drive and gantry synchrony.

Quality assurance programs for IGRT are not easy to implement. Rapid development of new IGRT techniques and devices is quickly making traditional QA guidelines outdated. Because of the diversity of IGRT, it is extremely difficult to develop industry-wide specific QA guidelines, forcing a conversion to process-centered quality management guidelines, which each institution can tailor to its individual needs. An optimal QA program is always a balance between available resources, manpower, and time to perform the work.

Patient-specific QA procedures for SRS/SBRT should be developed as an integrated part of a comprehensive ongoing QA program in the clinic. Therefore, before implementing an SBRT program, the clinic first needs to determine which system(s) will be used and develop QA procedures to match. SBRT-enabled systems often have specialized equipment such as immobilization systems, localization systems, and on-board imaging systems which are not always found in the clinic. In other cases, the entire system is specialized for SBRT (e.g. the Accuray Cyberknife). For example, Table 1 summarizes the stereotactic localization and image guidance strategies used by commercially-available systems. These specialized components require detailed and specialized QA procedures, over and beyond the general guidelines for external beam radiotherapy as specified in the AAPM Reports of TG 40, 142, and 45.

Author	Site	Immobilization/repositioning	Reported accuracy
Lax, 1994	Abdomen	Wood frame/stereotactic coordinates on box to skin marks	3.7 mm Lat. 5.7 mm Long
Hamilion, 1995	Spine	Screw fixation of spinous processes to box	2 mm
Murphy, 1997	Spine	Frameless/implanted fiducial markers with real-time imaging and tracking	1.6 mm radial
Lohr, 1999	Spine	Body cast with stereotactic coordinate	≤ 3.6 mm mean vector
Yenice, 2003	Spine	Custom stereotactic frame and in-room CT guidance	1.5 mm system accuracy, 2-3 mm positioning accuracy
Chang, 2004	Spine	MI™ BodyFix with stereotactic frame/linac/CT on rails with 6D robotic couch	1 mm system accuracy
Tokuuye, 1997	Liver	Prone position jaw and arm straps	5 mm
Nakagawa, 2000	Thoracic	MVCT on linac	Not reported
Wulf, 2000	Lung, liver	Elekta™ body frame	3.3 mm Lat. 4.4 mm long
Fuss, 2004	Lung, liver	MI™ Body Fix	Bony anatomy translation 0.4, 0.1, 1.6 mm (mean X, Y, Z); tumor translation before image guidance 2.9, 2.5, 3.2 mm (mean X, Y, Z)
Herfarth 2001	Liver	Leiginger body frame	1.8 – 4.4 mm
Nagata, 2002	Lung	Elekta™ body frame	2 mm
Fukumoto, 2002	Lung	Elekta™ body frame	Not reported
Hara, 2002	Lung	Custom bed transferred to treatment unit after confirmatory scan	2 mm
Hof, 2003	Lung	Leibinger body frame	1.8 – 4 mm
Timmerman, 2003	Lung	Elekta™ body frame	Approx. 5 mm
Wang, 2006	Lung	Medical Intelligence body frame stereotactic coordinates/CT on rails	0.3±1.8 mm AP. -1.8±3.2 mm Lat. 1.5±3.7 mm SI

Table 1. Achievable accuracies reported in the literature categorized by body site and immobilization/repositioning device.(AAPM TG101)

Source	Purpose	Proposed test	Reported achievable	Proposed frequency
Ryu et al., 2001	End-to-end localization accuracy	Stereo x ray/DRR fusion	1.0 to 1.2 mm root mean square	Initial commissioning and annually thereafter
Ryu et al., 2001	Intrafraction targeting variability	Stereo x ray/DRR fusion	0.2 mm average, 1.5 mm maximum	Daily (during treatment)
Verellen et al., 2003	End-to-end localization accuracy	Hidden target (using stereo x ray/DRR fusion)	0.41 ± 0.92 mm	Initial commissioning and annually thereafter
Verellen et al., 2003	End-to-end localization accuracy	Hidden target (using implanted fiducials)	0.28 ± 0.36 mm	Initial commissioning and annually thereafter
Yu et al., 2004	End-to-end localization accuracy	Dosimetric assessment of hidden target (using implanted fiducials)	0.68 ± 0.29 mm	Initial commissioning and annually thereafter
Sharpe et al., 2006	CBCT mechanical stability	Constancy comparison to MV imaging isocenter (using hidden targets)	0.50 ± 0.5 mm	Baseline at commissioning and monthly thereafter
Galvin et al., 2008	Overall positioning accuracy, including image registration (frame-based systems)	Wiston-Lutz test modified to make use of the in-room imaging system	≤ 2 mm for multiple couch angles	Initial commissioning and monthly thereafter
Palta et al., 2008	MLC accuracy	Light field, radiographic film or EPID	< 0.5 mm (especially for IMRT delivery)	Annually
Solberg et al., 2008	End-to-end localization accuracy	Hidden target in anthropomorphic phantom	1.0 ± 0.42 mm	Initial commissioning and annually thereafter
Jiang et al., 2008	Respiratory motion tracking and gating in 4D CT	Phantoms with cyclical motion	N/A	N/A
Bissonnette et al., 2008	CBCT geometric accuracy	Portal image vs CBCT image isocenter coincidence	2 mm	daily

Table 2. Summary of published QA recommendations for SBRT and SBRT-related techniques. .(AAPM TG101)

4. Clinical aspect

4.1. Proper selection of patients

The most important goal of SBRT in oligometastases is to achieve local control, however, whether obtaining local control of the metastasis would translate into clinical or survival benefit of the patients is dependent on multiple factors, including age, performance status, medical comorbidities and histology of malignancies. Therefore, the patients' whole condition should be fully considered. In general, patients with younger age, high performance status, controlled primary sites, limited number of metastases from three to five or fewer, metachronous occurrence of primary disease and metastatic disease, histologies, such as colorectal carcinoma, breast cancer and radioresistant cancer including renal cell ca, melanoma and sarcoma, are most likely to benefit from SBRT of their oligometastases (Carey-Sampson et al, 2006). In addition, SBRT delivers the individual ablative radiation doses to a planning target volume with a steep dose gradient outside the lesion treated and it is crucial that the lesions to be treated must be easily delineated on diagnostic imaging.

4.2. Lung metastases

	type	No of pts /targets	Dose(Gy/fx)	FU (mo)	LC (%)	Survival
Blomgren et al, 1995a	retrospective	10/14	7.7-45Gy/1-4fx	8	92	Med.S 11.3mo
Uematsu et al, 1995b	retrospective	22/43	33-71Gy/5-15fx	9	98	
Nakagawa et al, 2000	retrospective	14/21	16-24Gy/1fx	10	95	2YOS 35%
Wulf et al, 2001	retrospective	41/51	30-37.5 Gy/3fx ; 26 Gy/1fx	14	80%	2YOS 33%
Hara et al, 2002	retrospective	14/18	20-30Gy/1fx	12	78	
Lee et al, 2003	retrospective	19/25	30-40 Gy/3-4fx	18	88	Med.LPFS 18mo
Hof et al, 2007	retrospective	61/71	12-30 Gy/1fx	14	88.6 (1YR)	3YOS 47.8%
Okunieff et al, 2006	retrospective	42/125	50 Gy/10fx	18.7	94	Med.S 23.4mo
Norihisa et al, 2008	retrospective	34/43	48-60 Gy/4-5fx	27	90	2YOS 84.3%
Kim et al, 2009	retrospective	31/134	50 Gy/10fx	16	87.1	Med.S 16mo
Ernst-Stecken et al, 2006	prospective	21/39	35-40 Gy/5fx	NA	CR:51 PR:33 SD:3	Med.LPFS 6.4mo
Rusthoven et al, 2009	prospective	38/63	48-60 Gy/3fx	15.4	96	Med.S 19mo

Table 3. Results of SBRT in lung metastases

There are numerous retrospective studies on the use of SBRT for the treatment of lung oligometastases from North America, Europe and East Asia (Table 3). Early results from Blomgren's and Uematsu's studies showed excellent local control rates of 92% and 98%, respectively although the follow up periods were short (Blomgren et al, 1995; Uematsu et al, 1995). Subsequently, Nakagawa treated 14 patients with 21 tumors with SBRT to a single dose of 16 to 24 Gy. The local control rate and 2 year overall survival rate were 95% and 35%, respectively (Nakagawa et al, 2000). In a report of Wulf et al, the 41 patients with 51 metastatic lung tumors were treated with SBRT of 30 to 37.5 Gy in 3 fractions or 26 Gy of a single dose. The crude local control rate was 80% at a median follow up 14 months and 2 years overall survival rate was 33% (Wulf et al, 2001). And Hof et al also treated 61 patients with 71 lung metastases with SBRT to a single dose of 12 Gy to 30 Gy. The actuarial local progression free rate was 79% at 1 year and overall survival rate was 47.8% on 3 years (Hof et al, 2007). In a report of Okunieff et al, they treated 50 patients with five or fewer lung metastases with SBRT. At a median follow up of 18.7 months, 94% local control rate and 50% of 2 years overall survival rate were yielded (Okunieff et al, 2006). Kim et al also treated the patients with multiple lung metastases with SBRT to a dose of 50 Gy in 10 fractions during 2 weeks. The local control rate was 87.1% and median survival time was 16.0 months (Kim et al, 2009). Two prospective studies' outcomes were also shown in table 3. In a report from Germany, Ernst-Stecken et al reported the results of dose escalating phase I/II trial of SBRT for lung tumors, Overall, 21 patients (three with primary lung tumors) with 39 tumors were treated with SBRT starting at dose level of 35 Gy (7 Gy x 5) and the dose was then escalated to 40 Gy (8 Gy x 5). In total, 21 and 18 tumors were treated to 35 Gy and 40 Gy, respectively. Rates of complete response, partial response, stable disease and progressive disease were 51%, 33%, 3% and 13%, respectively (Ernst-Stecken et al, 2006). In 2009, in a multi-institutional phase I/II trial of SBRT for patients with 1 to 3 lung metastatic tumors less than 7 cm diameter, the total radiation dose was safely escalated from 48 Gy to 60 Gy in 3 fractions. The 2 year actuarial local control rate was 96% and median survival time was 19 months (Rusthoven et al, 2009).

4.3. Liver metastases

	type	No of pts /targets	Dose(Gy/fx)	FU (mo)	LC(%)	Survival
Blomgren et al, 1998	retrospective	17/21	20-40 Gy/1-2fx	9.6	95	
Katz et al, 2007	retrospective	69/174	30-55Gy/2-6fx	14.5	76-57	Med.S 14.5mo
Wulf et al, 2001	retrospective	23/23	28-30Gy/2-4fx	9	76-61	
Herfarth et al, 2001	prospective	33/56	14-26Gy/1fx	18	67	1YSR 72%
Kanavagh et al, 2006	prospective	21/28	36-60Gy/3fx	18	93	
Mendez-Romero et al, 2006	prospective	17/34	37.5Gy/3fx	12.9	100-86	

Table 4. Results of SBRT in liver metastases

Blomgren's early data on SBRT for liver metastases showed promising results of 95% local control rate on 9.6 months follow up (Blomgren et al, 1998). In the study of University of Rochester, which represents the largest study in SBRT for liver metastases, Katz et al treated 69 patients with 174 liver metastases with SBRT to a median dose of 48 Gy(range, 30-55Gy) in 2 to 6 fractions. The mean number of lesions was 2.5 (range, 1-6). The most common primary sites were colorectal (n=20) and breast (n=16). The median follow up was 14.5 months. The local control rates were 76% and 57% at 10 and 20 months, respectively. The median overall survival time was 14.5 months (Katz et al, 2007). Wulf et al reported their experience on 23 patients treated with SBRT for liver metastases. The prescribed dose was 30 Gy in three fractions. The actuarial local control rates on one and two year after treatment were 76 and 61%, respectively (Wulf et al, 2001). Herfarth et al performed a dose escalation study utilizing single dose SBRT from 14 Gy to 26 Gy. Fifty six liver metastases of 33 patients were treated and their local control rate was 67% on 18 months after treatment. Local failures were observed mainly in patients treated to a lower dose. For patients treated to higher dose (>20 Gy), the actuarial local control rate was 81% (Herfarth et al, 2001). In a study of Colorado University, Kavanagh et al reported 93% of actuarial local control rate on 18 months and indicated that a very high rate of durable in-filed tumor control can be safely achieved with SBRT to one to three liver lesions to a prescription dose of 60 Gy in 3 fractions (Kanavagh et al, 2006). Mendez-Romero et al reported the results of 17 patients with 34 metastatic liver tumors treated in phase I/II study of SBRT. The prescribed dose was 37.5 Gy in 3 fractions. The actuarial one and two year local control rates were 100% and 86%, respectively and the actuarial overall survival rate at one and two years were 85% and 62%, respectively (Mendez-Romero et al, 2006).

4.4. Spine metastases

SBRT has emerged as a novel treatment modality in the multidisciplinary management of spinal metastasis. Compared with conventional radiotherapy, SBRT can deliver a much higher biologic equivalent dose to the spinal tumor while respecting the dose constraints of the spinal cord or cauda equine, which are usually the dose limiting structures. The inclusion criteria for spinal SBRT are solitary or oligometastatic disease or bone only disease in otherwise high performance status patients, maximum of two consecutive or non contiguous spinal segments involved by tumor, failure of prior XRT (upto one course and 45 Gy maximum) or surgery, non myeloma tumor type, gross residual disease or deemed to high risk for recurrence postsurgery, patients refusal or medical comorbidities precluding surgery, gross tumor optimally more than 5 mm from the spinal cord, Karnofsky performance status > 40-50, MRI- or CT documented spinal tumor, histologic confirmation of neoplastic disease and Age > 18. These are yielded from reports by various authors for spine SBRT. And these criteria are based on relevant studies, which include those reporting both the dose/fractionation used and duration of follow up for patients treated for metastatic spinal tumors. However, the final treatment recommendation should involve ideally a multidisciplinary tumor board composed of surgeons, radiation oncologists, medical oncologists, and medical physicists.

	Number of Tumor/pts	Target volume/image	Dose/fx	Re-RTx	FU (mo)	LC/criteria
Ryu et al, 2004	61/49	Involved spinal segment/CT or MR	10-16 Gy/1fx	ERT 25 Gy/10 plus SBRS boost 6-8 Gy/1	6-24	93%/imaging and clinical
Milker-Zabel et al, 2003	19/18	PTV=GTV plus entire VB/CT-MRI fusion	24-45 Gy, Median 2 Gy fraction	19/18 Median 39.6 Gy, 2 Gy fraction	12	95%/clinical
Gerszten et al, 2005	26/26	Postkypoplasty VB+extension/ CT	16-20 Gy/1fx		4-36	92%/imaging or clinical
Gerszten et al, 2007	500/393	GTV=PTV/CT	12.5-25 Gy/1	7 patients combined EBRT plus SBRT boost	3-53	88%/imaging
Sahgal et al, 2007	60/38	GTV=PTV/CT	8-30 Gy/1-5	37/26 tumors had previous irradiated	1-48	87%/imaging and clinical
Chang et al, 2007	74/63	GTV + potential extension of structure /CT	30 Gy/5fx or 27 Gy/ 3fx	35/63 (55.6%) patients of previous spinal RT (median 33 Gy; range 30-54 Gy)was allowed	1-50	77%/imaging

Table 5. Clinical Results of SBRT in spinal metastasis

In a report from Henry Ford Hospital, Ryu et al treated 61 spinal tumors in 49 patients with single dose of SBRT alone to a dose of 10 to 16 Gy. With follow up time ranging from 6 to 24 months, the local control rate was 93% on imaging and clinical response including complete or partial pain control was achieved in 52 of 61 tumors (85%) (Ryu et al, 2004). In a report of SBRT as reirradiation, Milker-Zable treated 19 tumors from 18 patients with a dose range from 24 to 45 Gy in 2 Gy fractions. Their previous median dose was 39.6 Gy in 2 Gy fractions. With a median follow up time 12 months, the clinical response rate was 95%. They defined PTV as a gross tumor volume plus entire vertebral body through CT with MRI fusion and defined spinal cord as spinal cord from MRI plus safety margin of 2 to 3 mm. Dose constraints of spinal cord on SBRT as reirradiation was maximal dose to spinal cord less than 20 Gy in 10 fractions to a median percent of spinal cord (Milker-Zabel et al, 2003). In a postoperative SBRT series from Pittsburg Medical center, Gerszten et al reported the results of SBRT using Cyberkinife from 26 tumors in 26 patients. The prescribed dose was 16 to 20 Gy at the 80% isodose line with a median follow up of 16 months, the local control rate was 92%. Pain control was evaluated using a ten point verbal visual analog scale and was improved in 24 out of 26 patients (Gerszten et al, 2005). And in the largest report from same group, Gerszten et al treated a total of 393 patients with 500 spinal metastases with Cyberknife based single dose SBRT to doses ranging from 12.5 to 25 Gy. Seven patients also received external radiation therapy. With a median follow up of 21 months, the local control rate was 88%. Among the 336 evaluable patients, 290 (86%) achieved improvement in pain based on a ten point visual analog scale (Gerszten et al, 2007). Sahgal et al reported the treatment results of Cyberknife based SBRT for spinal metastases from University of California SanFrancisco in abstract form. They treated 60 spinal metaststases in 38 patients with a dose ranging from 8 to 30 Gy in one to five fractions (median 24 Gy in three fractions). With a median follow up of 8.5 months, the local control rate was 87% and the pain improvement was achieved in 31 out of 46 tumor sites (67%) (Sahgal et al, 2007). In a phase I/II trial from MD Anderson Cancer Center, Chang et al reported the results of 63 patients with 74 tumors treated with SBRT to a dose of 30 Gy in five fractions or 27 Gy in three fractions. Thirty five patients had prior external radiotherapy. With a median follow up of 21.3 months, the local control rate was 77% and the one year progression free rate was 84% (Chang et al, 2007).

4.5. Multiple organ oligometastases

Authors	No of pts/tumors	Sum of GTV	Dose/fx	FU (mo)	Outcome
Milano et al, 2008	121 /293	0.3-422 ml Med. 28 ml	50Gy/10 (SRS 10-20Gy/1)		2 year OS/PFS/LC/DC, 50%/26%/67%/34%; 4 year OS/PFS/LC/DC, 28%/20%/60%/25%
Salama et al, 2008	29/56	Max. dimension of volume ≤ 10cm or < 500cm³	24-36 Gy/3	5.3-27 (med. 14.9)	Response rate 59%; PFSR 21%; LC 57%
Salama et al, 2011	61/113	Max. dimension of volume ≤ 10cm or < 500cm³	24-48 Gy/3	Med. 20.9	1 year OS/PFS 81.5%/33.3%; 2 year OS/PFS 56.7%/22.0%

Table 6. Clinical results of 5 or fewer oligometastases

There are fewer reports about SBRT in multisite oliogometastases (Table 6). Among them the largest trial was performed in Rochester University hospital. Milano et al reported that the 4 year overall survival, progression free survival, local control and distant control were 28%, 20%, 60% and 25%, respectively after SBRT for multiple sites oligometastases from 121 patients. And they showed that number of metastases (range, 1~5) was not correlated with treatment outcomes. Salama et al firstly performed dose escalation study of SBRT in patients with oligometastases involving multiple organs (Milano et al, 2008). In phase I/II trial, they treated 56 tumors in 29 patients with a dose to 24 to 36 Gy in 3 fractions. With a median follow up 14.9 months, local control and progression free survival rate were, 57% and 21%, respectively (Salama et al, 2008). In a final report from same group, Salama et al could escalate the dose from 24 Gy to 48 Gy in 3 fractions. Fifty six tumors in 29 patients were treated and their 1 and 2 year overall survival rate was 81.5% and 56.7%, respectively. And they showed superior outcome in the patient with one to three metastases to the others with four or five metastases (2 year overall survival; 60.3% vs 21.9%) but there was not statistical significance (p=0.22) (Salama et al, 2011).

4.6. Dose constraints to prevent normal tissue toxicity

SBRT has been defined as hypofractionated (1-5 fractions) extracranial stereotactic radiation delivery, thus when selecting the fractional and total dose, several clinical considerations are important, including; (1) predicted risks of late normal tissue complications; (2) predicted tumor control; (3) financial costs and time expenditure for treatment planning and delivery. Among these, the long term effect of hypofractionated dose delivery to small volumes of normal tissues is not well understood, and certainly more clinical studies with longer follow up are needed to better define the variable associated with risks of late toxicity. Table 7 shows the normal tissue dose volume constraints to prevent late radiation complication in NCCN guidleline version 2.2012.

OAR	1 fraction	3 fractions	4 fractions	5 fractions
Spinal cord	14 Gy	18 Gy (6 Gy/fx)	26 Gy (6.5 Gy/fx)	30 Gy (6 Gy/fx)
Esophagus	15.4 Gy	30 Gy (10 Gy/fx)	30 Gy (7.5 Gy/fx)	32.5 Gy (6.5 Gy/fx)
Brachial plexus	17.5 Gy	21 Gy (7 Gy/fx)	27.2 Gy (6.8 Gy/fx)	30 Gy (6 Gy/fx)
Heart/pericardium	22 Gy	30 Gy (10 Gy/fx)	34 Gy (8.5 Gy/fx)	35 Gy (7 Gy/fx)
Great vessels	37 Gy	39 Gy (13 Gy/fx)	49 Gy (12.25 Gy/fx)	55 Gy (11 Gy/fx)
Trachea & proximal bronchi	20.2 Gy	30 Gy (10 Gy/fx)	34.8 Gy (8.7 Gy.fx)	32.5 Gy (6.5 Gy/fx)
Rib	30 Gy	30 Gy (10 Gy/fx)	34.8 Gy (8.7 Gy.fx)	32.5 Gy (6.5 Gy/fx)
Skin	26 Gy	30 Gy (10 Gy/fx)	36 Gy (9 Gy/fx)	40 Gy (8 Gy/fx)
Stomach	12.4 Gy	27 Gy (9 Gy/fx)	30 Gy (7.5 Gy)	35 Gy (7 Gy/fx)

Table 7. Normal tissue dose volume constraints for SBRT from NCCN guidelines

The recommendation from Table 7 is frequently referenced in SBRT for non small cell lung cancer (Ettinger et al, 2012) and so, there is no information about intra abdominal organ including small intestine, liver and kidney. Radiobiologically, normal tissues can be categorized into two groups of serially arranged tissues and parallel arranged tissues. In a review article from Rochester University in New York, Milano et al recommended the fractional dose limitations to small volume of normal tissue which were expected to be safe with respect to risk of radiation necrosis in serially arranged tissues and they also noted the dose constraints of parallel arranged normal tissues such as lung, liver and kidney for safe SBRT in same article (table 8 and 9) (Milano, 2008).

	Number of fractions				
Normal tissue	1	3	5	8	10
Spinal cord	8-10 Gy	5-6 Gy	4-5 Gy	3-4 Gy	3 Gy
Trachea & bronchi	-	-	7-9 Gy	6-7 Gy	4-5 Gy
Brachial plexus	-	-	8-10 Gy	6-7 Gy	5-6 Gy
esophagus	-	-	6-8 Gy	4-5 Gy	3-4 Gy
Chest wall/ribs	-	10-15 Gy	6-8 Gy	6-7 Gy	5-6 Gy
Small bowel	10-12 Gy	10-12 Gy	6-8 Gy	5-6 Gy	4-5 Gy
Lung	20 Gy	20 Gy	8-10 Gy	7-8 Gy	5-7 Gy
Liver	25 Gy	20 Gy	8-10 Gy	7-8 Gy	5-6 Gy

Table 8. Recommendation for safe hypofractionated SBRT fractional dose to small volume of serially arranged tissues.

Lung	700-1000 ml of lung not involved with gross disease or planning target volume V20 of 25-30%
Liver	700-1000 ml of liver not involved with gross disease or planning target volume Two thirds of normal liver < 30 Gy
Kidney	Minimize dose receiving > 20 Gy Two thirds of one kidney < 15 Gy (with another functional kidney)

Table 9. Recommendation for safe hypofractionated SBRT dose volume metrics for parallel arranged normal tissues

Deriving standard acceptable maximally effective and minimally toxic dose fractionation schemes presents a challenge, even with available outcome data. In fact, this complexity arises from not only the different dose-fractionation schemes used, but also in differences in how the dose is prescribed. Further study and longer follow up are needed to ascertain the dose fractionation schedule that optimizes tumor control while minimizing toxicity and to better understand the optimal normal tissue dose volume constraints.

4.7. Patterns of failure

A subset of patients with oligometastases have been alive a prolonged disease free state, some > 7 years, most eventually succumbed to further metastatic progression. There are several studies which have examined the pattern of recurrence after resection, radiofrequency ablation, or cryosurgery and SBRT of oligometastases. Table 10 shows the literature summary of the pattern of recurrence after treatment of limited liver metastases.

First author	Sugihara et al, 1993	Aloia et al, 2006	Kosari et al, 2002	Ravikumar et al, 1991	Milano et al, 2010
Primary cancer	colorectal	colorectal	various	colorectal	various
Treatment modality	resection	resection	radio-frequency	cryosurgery	SBRT
Number of recur/total	64/107 (60%)	71/150 (57%)	23/45 (51%)	17/24 (71%)	37/42 (88%)
Follow up(mo)	6-164 Median 35	4-138 Median 31	6-34 Median 19.5	5-60 Median 24	6-67 Median 21
Recurrence in					
Liver only	-	18%	52%	35%	22%
Extrahepatic only	-	62%	4%	6%	5%
Liver+extrahepatic	-	20%	43%	59%	73%
Liver	53%	38%	96%	94%	95%
Lung	31%	58%	-	-	32%
CNS	-	1%	-	-	8%
Bone	-	6%	-	-	19%
other	28%	17%	-	-	32%

Table 10. The pattern of recurrence after local treatment of limited liver metastases

All authors reported that the first new recurrence or metastases occurred quite commonly in the same organ, although metastases to other organs are common as well. New metastases occurring shortly after completion of treatment including SBRT presumably represents the growth of initially occult metastatic disease versus rapid metastatic progression, whereas new metastases that occurs after a longer time interval represents more indolent growth of initially occult metastatic disease versus a more remote occurrence of distant spread. However, a few present studies can determine a mechanism to account for new metastases. Some variables are thought important in predicting where subsequent metastases are likely to occur. The initial organ involvement, use of chemotherapy, type of local therapy, primary cancer type, histology and grade are expected to be important which can impact the pattern of subsequent recurrence. In addition, genotypic and phenotypic changes which lead to metastatic potential must exist and should be explained in the future.

5. Conclusion and future aspect

In its current form, stereotactic hypofractionated radiotherapy is still in its infancy as an experimental treatment for oligometastases. At this point, a recommendation cannot be made for a fractionation scheme, which suggests the need for prospective investigation. There are multiple ongoing clinical trials on the use of SBRT for oligometastases in various body sites and the results of those trials are eagerly awaited. Given the high propensity for distant progression, the combination of novel systemic therapy and SBRT is to be explored. Interested readers can visit the web site (www.clinicaltrials.gov) to a full list of clinical trials of SBRT for various metastatic sites.

Author details

Chul-Seung Kay
The Catholic University of Korea, Republic of Korea,
Incheon St. Mary Hospital Republic of Korea

Young-Nam Kang
The Catholic University of Korea, Republic of Korea

6. References

Aloia TA, Vauthey JN & Loyer EM (2006) Solitary colorectal liver metastasis: resection determines outcome. *Arch Surg* 141, 5 (May, 2006) 460-466, ISSN 0004-0010

Andrews DW, Scott CB & Sperduto PW (2004) Whole brain radiation therapy with or without stereotactic radiosurgery boost for patients with one to three brain metastases: Phase III results of the RTOG9508 randomized trial. *The Lancet* 363, 9422, (May 2004) 1665–1672, ISSN 1474-547X

Argiris A, Li Y & Forastiere A (2004) Prognostic factors and long term survivorship in patients with recurrent or metastatic carcinoma of the head and neck. *Cancer,* 101, 10 (November 2004), 2222-2229, ISSN 1097-0142

Benedict SH, Yenice KM & Followill DF (2010). Stereotactic body radiation therapy: The report of AAPM Task Group 101. *Med. Phys.* 37, 8, (August 2010) ISSN 0094-2405

Blomgren H, Lax I & Noslund I (1995) Stereotactic high dose fraction radiation therapy of extracranial tumors using and accelerator. Clinical experience of the first thirty one patients. *Acta Oncol,* 34, 6 (1995), 861-870, ISSN 0284-186X

Blomgren H, Lax I & Goranso H (1998) Radiosurgery for tumors in the body: clinical experience using a new method. *J Radiosurg,* 1, 1, (December, 1998), 63-74, ISSN 1096-4053

Brenner DJ (2008) The linear quadratic model is an appropriate methodology for determining isoeffective dose at large doses per fraction. *Sem Radiat Oncol,* 18, 4 (October, 2008), 234-239, ISSN 1053-4296

Carey-Sampson M, Katz AW & Constine LS (2006) Stereotactic body radiation therapy for extracranial oligometastasis: Does the sword have a double edge? *Semin Radiat Oncol,* 16, 2 (April, 2006) 67-76, ISSN 1053-4296

Carmichael J, Degraff WG & Gamson J (1989) Radiation sensitivity of human lung cancer cell lines. *Eur J Cancer Clin Oncol,* 25, 3 (March, 1989), 527-534, ISSN 0277-5379

Chang EL, Shiu AS & Mendel E (2007) Phase I/II study of stereotactic body radiotherapy for spinal metastases and its pattern of failure. *J Neurosurg Spine,* 7, 2 (August, 2007), 151-160, ISSN 1547-5654

Ettinger DS, Akerly W & Borghaei H (2012) NCCN clinical practice guidelines in oncology (NCCN guidelines®) Non small cell lung cancer version 2.2012 In: *National Comprehensive Cancer Network,* available from http://www.nccn.org/professionals/physician_gls/pdf/nscl.pdf

Ernst-Stecken A, Lambrecht U & Mueller R (2006) Hypofractionated stereotactic radiotherapy for primary and secondary intrapulmonary tumors: first results of a phase I/II study. *Strahlenther Onkol,* 182, 12 (December, 2006), 696-702, ISSN 0179-7158

Escudier B, Eisen T & Stadler WM (2007) Sorafenib in advanced clear cell renal cell carcinoma. *N Engl J Med,* 356, 2 (January, 2007), 125-134, ISSN 0028-4793

Flickinger JC, Kondziolka D & Niranjan A (2001) Results of acoustic neuroma radiosurgery: an analysis of 5 Years' experience using current methods. *J. Neurosurg.* 94, 1 (January 2001) 1–6, ISSN 0022-3085

Flickinger JC, Konziolka D & Maitz AH (2002) An analysis of the dose response fro arteriovenous malformation radiosurgery and other factors affecting obliteration. *Radio. Oncol,* 63, (December 2000) 347-354, ISSN 1748-717X

Fong Y, Fortner J & Sun RL (1999) Clinical score for predicting recurrence after hepatic resection for metastatic colorectal cancer: analysis of 1001 consecutive cases. *Ann Surg,* 230, 3 (September, 1999), 309-318, ISSN 0003-4932

Friedel G, Linder A & Toomes H (1994) The significance of prognositc factors for the resection of pulmonary metastases of breast cancer. *Thoac Cardiovasc Surg,* 42, 2 (April, 1994), 71-75, ISSN 0171-6425

Gerszten PC, Germanwala A & Burton SA (2005) Combination kyphoplasty and spinal radiosurgery: a new treatment paradigm for pathological fractures. *J Neurosurg Spine,* 3, 4 (October, 2005), 296-301, ISSN 1547-5654

Gerszten PC, Burton SA & Ozhasoglu C (2007) Radiosurgery for spinal metastases: clinical experience in 500 cases from a single institution. *Spine,* 32, 2 (January, 2007), 193-199, ISSN 0362-2436

Guerrero M & Li XA (2004) Extending the linear quadratic model for large fraction doses pertinent to stereotactic radiotherapy. *Phys Med Biol,* 49, 20 (October, 2004), 4825-4835, ISSN 0031-9155

Hall EJ (1993) The radiobiology of radiosurgery: Rationale for different treatment regimens for AVM and malignancies. *Int J Radiat Oncol Biol Phys,* 25, 2 (January , 1993), 381, ISSN 0360-3016

Hall EJ & Giaccia AJ (2006) Cell survival curves, In: *Radiobiology for the radiologist, 6th ed,* 30-46, Lippincott Williams & Wilkins, ISBN 0-7817-4151-3, Philadelphia, PA 19106, USA

Hara R, Itami J & Kondo T (2002) Stereotactic single high dose irradiation of lung tumors under respiratory gating. *Radiother Oncol,* 63, 2 (May, 2002), 159-163, ISSN 0167-8140

Hellman S & Weichselbaum RR (1995) Oligometastasis. *J Clin Onco,* 13, 1 (January, 1995), 8-10, ISSN 0732-183X

Herfarth KK, Debus J & Lohr F (2001) Stereotactic single dose radiation therapy of live tumors: results of a phase I/II trial. *J Clin Oncol,* 19, 1 (January, 2001), 164-170, ISSN 0732-183X

Hof H, Hoess A & Oetzel D (2007) Stereotactic single dose radiotherapy of lung metastases. *Strahlenther Onkol,* 183, 12 (December, 2007), 673-678, ISSN 0179-7158

Hurwits H, Fehrenbacher L & Novotony W (2004) Bevacizumab plus irinotecan, fluorouracil, and leucovorin for metastatic colorectal cancer. *N Engl J Med,* 350, 23 (June, 2004), 2335-2342, ISSN 0028-4793

Izawa M, Hayashi M & Nagata K (2000) Gamma Knife Radiosurgery for Ptuitay adenomas. *J Neurosurgery* 93, 1 (Decmber 2000), 19-22, ISSN 0022-3085

Katz AW, Carey-Sampson M & Muhs AG (2007) Hypofractionated stereotactic body radiation therapy for limited hepatic metastases. *Int J Radiat Oncol Biol Phys,* 67, 3 (March, 2007), 793, ISSN 0360-3016

Kavanagh BD, Schefter TE & Cardenes HR (2006) Interim analysis of a prospective phase I/II trial of SBRT for liver metastases. *Acta Oncol,* 45, 7 (2006), 848-855, ISSN 0284-186X

Kim JY, Kay CS & Kim YS (2009) Helical tomotherapy for simultaneous multitarget radiotherapy for pulmonary metastasis. *Int J Radiat Oncol Biol Phys,* 75, 3 (November, 2009), 703-710, ISSN 0360-3016

Kirkpatrick JP, Meuer JJ & Marks LB (2008) The linear quadratic model is inappropriate to model high dose per fraction effect in radiosurgery. *Sem Radiat Oncol,* 18, 4 (October, 2008), 240-243, ISSN 1053-4296

Konski AA Wallner PE & Harris EER (2008) Stereotactic body radiotherapy (SBRT) for lung cancer: Report of the ASTRO Emerging Technology Committee (ETC)

Kosari K, Gomes G & Hunter D (2002) Local, intrahepatic, and systemic recurrence patterns after radiofrequency ablation of hepatic malignancies. *J Gastrointest Surg,* 6, 2 (March-April, 2002), 255-263, ISSN 1091-255X

Lee SW, Choi EK & Park HJ (2003) Stereotactic body frame based fractionated radiosurgery on consecutive days for primary or metastatic tumors in the lung. *Lung cancer,* 40, 3 (June, 2003), 309-315, ISSN 0169-5002

MacDermed DM, Wechselbaum RR & Salama JK (2008) A rationale for the targeted treatment of oligometastases with radiotherapy. *J Surg Oncol,* 98, 3 (September 2008), 202-206, ISSN 0022-4790

Mehta N, Mauer AM & Hellman S (2004) Analysis of further disease progression in metastatic non small cell lung cancer: implication for locoregional treatment. *Int J Oncol 25*, 6 (December, 2004), 1677-1683, ISSN 1019-6439

Mendez Romero A, Wunderlink W & Hussain SM (2006) Stereotactic body radiation therapy for primary and metastatic liver tumors: a single institution phase I-II study. *Acta Oncol,* 45, 7 (2006), 831-837, ISSN 0284-186X

Meyer JL, Kavanagh BD & Purdy JA (2007). *IMRT.IGRT.SBRT Advances in the Treatment Planning and Delivery of Radiotherapy*. KARGER, ISBN 978-3-8055-8199-8

Milker-Zabel S, Zabel A & Thilmann C (2003) Clinical results of retreatment of vertebral bone metastases by stereotactic conformal radiotherapy and intensity modulated radiotherapy. *Int J Radiat Oncol Biol Phys,* 55, 1 (January, 2003), 162-167, ISSN 0360-3016

Milano MT, Katz AW & Muhs AG et al (2008) A prospective pilot study of curative intent stereotactic body radiation therapy in patients with 5 or fewer oligometastatic lesions. *Cancer,* 112, 3 (February, 2008), 650-358, ISSN 0008-543X

Milano MT, Constine LS & Okunieff P (2008) Normal tissue toxicity after small field hypofractionated stereotactic body radiation. *Radiat Oncol,* 3, 36 (October, 2008), doi:10.1186/1748-717X-3-36, ISSN 1748-717X

Milano MT, Katz AW & Okunieff P (2010) Patterns of recurrence after curative intent radiation for oligometastases confined to one organ. *Am J Clin Oncol,* 33, 2 (April, 2010), 157-163, ISSN 0277-3732

Morstyn G, Russo A & Carney DN (1984) Heterogeneity in the radiation survival curves and biochemical properties of human lung cancer cell lines. *J Natl Cancer Inst,* 73, 4 (November, 1984), 801-807, ISSN 0027-8874

Nakagawa K, Aoki Y & Tago M (2000) Megavoltage CT-assisted stereotactic radiosurgery for thoracic tumors: original research in the treatment of thoracic neoplasm. *Int J Radiat Oncol Biol Phys,* 48, 2 (September, 2000), 449-457, ISSN 0360-3016

Norihisa Y, Nagata Y & Takayama K (2008) Stereotacic body radiotherapy for oligometastatic lung tumors. *Int J Radiat Oncol Biol Phys,* 72, 2 (April, 2008), 398-403, ISSN 0360-3016

Okunieff P, Peterson AL & Philip A (2006) Stereotatic body radiation therapy (SBRT) for lung metastases. *Acta Oncol,* 45, 7 (2006), 808-817, ISSN 0284-186X

Paget S (1889) The distribution of secondary growth in cancers in breast. *Lancet,* 8, 2 (August, 1889), 98, ISSN 0140-6736

Park C, Papiez L & Zhang S (2008) Universal survival curve and single fraction equivalent dose: useful tools in understanding potency of ablative radiotherapy. *Int J Radiat Oncol Biol Phys,* 70, 3 (March, 2008), 847-852, ISSN 0360-3016

Pastorino U, Buyse M & Friedel G (1997) Long term results of lung metastasectomy: prognostic analyses based on 5206 cases. *J Thorac Cardiovasc Surg,* 113, 1 (January, 1997), 37-49, ISSN 0022-5223

Pocard M, Pouilart P & Asselain B (2001) Hepatic resection for breast cancer metastases: results and prognosis. *Ann Chir,* 126, 5 (June, 2001), 413-420, ISSN 0753-9503

Pollock BE, Phuong LK & Gorman DA (2011) Stereotactic Radiosurgery for Idiopathic trigerminal Neuralgia. *J Neurosurg* 115, 1, (August 2002), 347-353, ISSN 0022-3085

Ravikumar TS, Steel G Jr & Kane R (1991) Experimental and clinical observations on hepatic cryosurgery for colorectal metastasis. *Cancer Res,* 51, 23 (December, 1991), 6323-6327, ISSN 0008-5472

Rusthoven KE, Hammerman SF & Kanavagh BD et al (2009) Is there a role for consolidative stereotactic body radiation therapy following first line systemic therapy for metastatic lung cancer? A patterns-of-failure analysis. *Acta Oncol,* 48, 4 (2009), 578-583, ISSN 0284-186X

Rusthoven KE , Kanavagh BD & Burri SH (2009) Multi-institutional phase I/II trial of stereotactic body radiotherapy for lung metastases. *J Clin Oncol,* 27, 10 (April, 2009), 1579-1584, ISSN 0732-183X

Ryu S, Rock J & Rosenblum (2004) Pattern of failure after single dose radiosurgery for spinal metastasis. *J Neurosurg,* 101, suppl 3 (November, 2004), 402-405, ISSN 0022-3085

Sahgal A, Larson DA & Chang EL (2008) Stereotactic body radiosurgery for spinal metastases: a critical review. *Int J Radiat Oncolo Biol Phys,* 71, 3 (July, 2008), 652-665, ISSN 0360-3016

Sahgal A (2007) Proximity of spinous/paraspinous radiosurgery metastatic targets to the spinal cord versus risk of local failure. *Int J Radiat Oncol Biol Phys,* 69, 3 (November, 2007), S243, ISSN 0360-3016

Salama JK, Chmura SJ & Mehta N (2008) An initial report of a radiation dose escalation trials in patients with one to five sites of metastatic disease. *Clin Cancer Res,* 14, 16 (August, 2008), ISSN 1078-0432

Salama Jk, Hasselle MD & Chmura SJ (2011) *Cancer,* Epub ahead of print (October, 2011) doi:10.1002/cncr.26611

Stafford SL, Pollock BE & Foote RL (2001) Meningioma Radiosurgery: tumor contro, outcomes and complications among 190 consecutive patiens. Neurosurgery 49, 5, (November 2001), 1029-1038, ISSN 1524-4040

Sugihara K (1993) Pattern of recurrence after hepatic resection for colorectal metastases. *Br J Surg,* 80, 8 (August, 1993), 1032, ISSN 0007-1323

Timmerman RD, Kavanagh BD & Cho LC (2007) Stereotactic body radiation therapy in multiple organ sites. *J Clin Oncol,* 25, 8 (March, 2007), 947-952, ISSN 0732-183X

Uemetsu M, Shioda A & Tahara K (1998) Focal, high dose and fractionated modified stereotactic radiation therapy for lung carcinoma patients. *Cancer,* 82, 6 (March, 1998), 1062-1070, ISSN 1097-0142

Underberg RWM, Lagerwaard FJ & Cuijpers JP (2004). Four-Dimensional CT Scans for Treatment Planning in Stereotactic Radiotherapy for Stage I Lung Cancer. *Int. J. Radia. Oncol. Biol. Phys.* 60, 4 (July 2004)), 1283-1290, ISSN 0360-3016

Wulf J, Haedinger U & Oppitz U (2001) Stereotactic radiotherapy of targets in the lung and liver. *Strahlenther Onkol,* 177, 12 (December, 2001), 645-655, ISSN 0179-7158

Wulf J, Haedinger U & Oppitz U (2004) Stereotactic Radiotherapy for primary lung cancer and pulmonary metastases: a noninvasive treatment approach in medically inoperable patients. *Int J Radiat Oncol Biol Phys,* 60, 1 (September 2004), 186-196, ISSN 0360-3016

Chapter 8

Locally Advanced Esophageal Cancer

Hend Ahmed El-Hadaad and Hanan Ahmed Wahba

Additional information is available at the end of the chapter

http://dx.doi.org/10.5772/56484

1. Introduction

Cancer of the esophagus is a highly lethal malignancy. There are approximately 16,980 people diagnosed with esophageal cancer each year in the United States and 14,710 deaths from the disease (Siegel et al., 2011). It currently ranks ninth among the most frequent cancers in the world (Lerut et al., 2001), and it is the sixth leading cause of death from cancer (Falk et al., 2007). Although the best treatment for locally advanced esophegeal cancer is still being debated, the use of neoadjuvant chemoradiotherapy has gained acceptance (Tepper et al., 2008). The rationale for chemoradiotherapy (CRT) followed by surgery has potential to downsize the tumor, thereby increasing the rate of tumor-free (RO) resections, reducing early relapses, and improving survival (Swisher et al., 2005; Brucher et al.,2006). Chemoradiotherapy (CRT) has proved effective against resectable/unresectable esophageal squamous cell carcinoma. The Radiation Therapy Oncology Group (RTOG) trial 85-01 demonstrated the superiority of CRT with cisplatin (CDDP), 5-fluorouracil (5-FU), and concurrent irradiation (50.4 Gy) over radiotherapy alone (64 Gy) in patients with T1–3N0–1M0 esophageal cancer (Herskovic et al., 1992). Definitive chemoradiotherapy is appropriate for locally advanced cancer in patients who do not want surgery or in whom surgery is not possible as a result of technical or medical reasons. The higher doses of radiation administered with concurrent chemotherapy was explored in the protocol RTOG9504 which established 50.4 Gy as the standard dose of radiation to be administered concurrently with chemotherapy (Minsky et al., 2002). Three-dimensional conformal radiotherapy (3D-CRT) is an approach to the planning and delivery of radiation therapy and numerous investigators have demonstrated the benefits of this modality in a variety of cancers. These benefits include its normal tissue-sparing capabilities and its ability to deliver higher radiation doses compared with conventional radiotherapy (Oh et al., 1999).

To enhance the efficacy and tolerability of multimodal treatment, new chemotherapeutic agents such as oxaliplatin and capecitabine have been incorporated into esophegeal cancer therapy. Oxaliplatin is a third generation platinum compound, it forms inter and intrastrand cross links with DNA that inhibit DNA replication and transcription. In phase I and II, trials suggest that it has been found to be at least as effective as cisplatin in esophageal cancer and better tolerated (Khushalani et al., 2002). Neoadjuvant concurrent chemoradiotherapy with capecitabine and oxaliplatin concurrently with conformal radiotherapy in patients with locally advanced esophageal cancer better tolerated and effective with OAR 54.8% (Wahba et al., 2012).

Objectives:

The chapter examines:

- Diagnostic procedures
- Recommendations for treatment of locally advanced esophageal cancer
- The use of neoadjuvant chemotherapy and radiotherapy for treatment of locally advanced esophageal cancer
- Concurrent chemotherapy and radiotherapy for definitive treatment
- Follow-up care

2. Diagnosis and staging

In Western countries, the diagnosis of esophageal cancer is generally made by endoscopic biopsy of the esophagus. In the Far East, cytologic evaluation is frequently used.

The most accurate staging modalities are CT scanning and endoluminal ultrasound (EUS). CT scanning most accurately detects distant visceral metastases although both EUS and laparoscopic ultrasound are capable of detecting small metastases, particularly in the left lobe of the liver that may be missed on CT (Nguyen et al., 1999 & Wakelin et al., 2000). In locoregional staging, EUS is considerably more accurate than CT.

PET: Positron emission tomography (PET) scanning is a more recently described staging modality that detects uptake of fluorodeoxyglucose by tumor cells. Early studies suggest it may be more reliable than EUS alone in detecting nodal metastases (Choi et al., 2000 & Lerut et al., 2000), several recent studies have applied PET scanning for the assessment of response to neoadjuvant chemoradiation, demonstrating a correlation between fluorodeoxyglucose uptake, pathologic response at surgery, and subsequent survival (Kato et al., 2002).

As PET becomes more widely available; its use will probably become an important part of the preoperative evaluation of these patients.

Accurate staging provides useful information relating to prognosis and has considerable therapeutic implications.

3. Treatment of locally advanced esophageal cancer

3.1. Treatment principles

The treatment of locally advanced esophageal cancer is a multidisciplinary approach; single modality approaches have disappointing control rates.

Radiochemotherapy is the standard of care; neoadjuvant chemoradiotherapy is associated with a higher response rate in comparison to chemotherapy alone.

Definitive chemoradiation remains a reasonable therapeutic option for patients.

External beam radiation therapy alone can be considered for definitive treatment when chemotherapy is contraindicated.

For palliation of symptomatic locally advanced esophageal cancer, radiotherapy is highly effective, also endoscopic procedures such as dilating, stenting, and laser ablative techniques are effective in rapid symptoms alleviation.

3.2. Neoadjuvant chemotherapy

A trial performed by the Medical Research Council Oesophageal Cancer Working Group randomized 802 patients to surgery alone versus two cycles of preoperative cisplatin/5-FU. At a relatively short median follow-up of only 2 years, the chemotherapy-treated group demonstrated improved median OS (16.8 vs 13.3 months) and 2-year survival (43% vs 34%). The curative resection rate was improved marginally from 55% to 60%, and the pCR rate was 4% in the preoperative therapy group (Medical Research Council Oesophageal Cancer Working Group 2002). A French trial of 224 patients with gastric or lower esophageal adenocarcinoma (Boige et al., 2007), in which patients were randomized to two or three cycles of preoperative cisplatin/5-FU followed by surgery versus surgery alone. Those patients who appeared to benefit clinically or radiographically from preoperative therapy or who had persistent T3 or node-positive disease at surgery also received an additional three or four cycles of chemotherapy. Preoperative chemotherapy was associated with a significant improvement in R0 resection rate (74% vs 87%), 5-year disease-free survival (34% vs 21%), and 5-year OS (38% vs 24%). the survival benefit seen with preoperative cisplatin/5-FU on this trial appears to be very similar to that seen with perioperative ECF in the MAGIC trial (Cunningham et al., 2006). Polee et al have evaluated a biweekly combination of cisplatin and paclitaxel in a phase II study, with promising results. Objective responses occurred in 59% of 49 patients. No patients had progressive disease. Although 71% of patients had severe neutropenia, it was often asymptomatic. Forty-seven patients underwent resection subsequently. Complete pathologic responses occurred in 14% of patients. The median survival of patients in this study was 20 months, but it was 32 months in patients who had disease responsive to chemotherapy. The 3-year survival rate was 32% (Polee et al., 2003).

3.3. Neoadjuvant RadiationTherapy

Trials that evaluated the use of preoperative radiation as a single modality have consistently reported no benefit. Whenever a survival benefit was suggested, it tends to be modest, similar to neoadjuvant chemotherapy alone (Arnott et al., 1992; 2005 & Nygaard et al., 1992). One randomized trial revealed no benefit for either preoperative radiation or chemotherapy, concluding that both treatment modalities might be necessary to treat both local and systemic disease. Also, a meta-analysis could not demonstrate a significant survival benefit for preoperative radiation as a single modality (Arnott et al., 1992).

3.4. Neoadjuvant chemoradiotherapy

Chemoradiotherapy typically involves regimens of cisplatin or mitomycin and continuous-infusion 5-FU, with radiotherapy dosages from 30 to 40 Gy and up to 60 Gy in some trials. It results in pCR rates of 20–40%, with long-term survival of no more than 25–35% (Coia et al., 1991 & Valerdi et al., 1993). Superior survival is consistently achieved, though, in patients achieving a pCR to chemoradiotherapy (up to 50–60% at 5 years) (Berger et al., 2005; Makary et al., 2003; Stahl et al., 2005 &Heath et al., 2000). A meta-analysis of randomized trial comparing neoadjuvant chemoradiation therapy followed by surgery with surgery alone found that neoadjuvant concurrent chemoradiation therapy improved 3-year survival (odds ratio, 0.66) compared with surgery alone, with a non significant trend toward increased mortality with neoadjuvant treatment (Kaklamanos et al., 2003).

Newer chemotherapy agents are active and may improve outcome over conventional cisplatin/5-FU-based regimen such as paclitaxel, irinotican, oxaliplatin, xeloda and docetaxel.

A phase II trial of 129 patients employed paclitaxel/carboplatin [Paraplatin]/5-FU with 45 Gy of radiation therapy followed by esophagectomy. A pathologic complete response was seen in 38% of patients, with a median survival of 22 months and a 3-year survival of 41% (Meluch et al., 2003).

Ajani et al reported a series of 43 patients who received 12 weeks of cisplatin and irinotecan (Camptosar) followed by weekly paclitaxel with infusional 5-FU and concurrent radiation therapy (4,500 cGy) and then esophagectomy. Therapy was well tolerated, with no deaths from chemotherapy or chemoradiation therapy, and an operative mortality rate of 5%. Cisplatin and irinotecan induced responses in 37% of patients, and 91% of patients underwent complete resection. Pathologic complete responses occurred in 26% of patients, and some tumor shrinkage was noted in 63% of patients. With a median follow-up of more than 30 months, the median disease progression free survival was 10. 2 months, the median survival was 22.1 months, and the 2-year survival was 42%. The patients who had a pathologic response to therapy had significantly better outcomes than the rest of the study population. However, systemic recurrences remained a prominent cause of failure, with five patients experiencing recurrence first in the brain and an additional five patients, in the liver (Ajani et al., 2004).

Neoadjuvant concurrent capecitabine and oxaliplatin with conformal radiotherapy (45Gy) in 42 patients reported OAR 54.8% and pathological response 38%.Median survival time was 20 months and 2-year survival rate 42 % (Wahba et al., 2012).

A phase II trial assessed the feasibility and safety of induction chemotherapy with cisplatin (25 mg/m^2 d1-5, d29-34)/docetaxel (75 mg/m^2 d1, d29)/5-fluorouracil ((5-FU, 750 mg/m^2 d1-5, d 29-34) followed by external beam radiotherapy concurrent with docetaxel (15 mg/m^2

d1,8,15,22) and 5-FU (300 mg/m^2 continuous infusion on the days of radiotherapy).Twenty-four patients with locally advanced carcinoma of the esophagus were included. Following chemotherapy and chemoradiation eligible patients underwent esophagectomy. Sixteen patients underwent resection. Pathologic complete remission was achieved in 5 of those 16 patients, 13 patients had downstaging of disease. R0 resection was feasible in all 16 patients (Eisterer et al., 2011).

The incidence of residual disease in patients who have a complete clinical response to chemoradiation therapy is 40%-50%. Patients with complete response following chemoradiation therapy have the best survival rates with surgery. RTOG 9207 Phase I/II treated 49 patients with concurrent 5-FU,cisplatin+radiotherapy (50Gy/25 fractions and high dose rate brachytherapy 5Gyx3 or low dose rate 20x1); reported 24% grade 4 toxicity,12% fistula,10% treatment related deaths with median survival 11 months. Brachytherapy not recommended due to high toxicity (Caspar et al., 2000).

3.5. Positron emission tomography-directed therapy

18F-2-fluoro-deoxy-D-glucose positron emission tomo-graphy (PET) scanning is emerging as an important tool to investigate response to therapy. Several studies have demonstrated that the degree of response detected by PET following preoperative chemoradiotherapy (Downey et al., 2003 & Flamen et al 2002) or chemotherapy (Ott et al., 2006 & Weber et al., 2001) is highly correlated with pathologic response at surgery and with patient survival.

The German MUNICON trial evaluated the strategy of taking patients with locally advanced GE junction tumors with a suboptimal response to 2 weeks of induction chemotherapy with cisplatin/5-FU, as determined by serial PET scans, directly to immediate surgery, instead of continuing with presumably ineffective chemotherapy (Lordic et al., 2007). Patients with a metabolic response by PET (defined as ≥35% reduction in standard uptake value between baseline and repeat PET scan) continued with an additional 12 weeks of chemotherapy prior to surgery. This trial revealed a significantly improved R0 resection rate (96% vs. 74%), major pathologic response rate (58% vs. 0%), median event-free survival (29.7 vs. 14.1 months), and median OS (median not reached vs. 25.8 months) for PET responders versus PET non responders. The outcome for PET non responders referred for immediate surgery was similar to the outcome of such patients in an earlier trial who completed 3 months of preoperative chemotherapy (Ott et al., 2006), indicating that non responding patients were not compromised by referral to immediate surgery.

3.6. Primary chemoradiation therapy

Patients with locally advanced esophageal cancer (T1-4 N0-1 M0) may be cured with definitive chemoradiation therapy. Randomized trials have demonstrated a survival advantage for chemoradiation therapy over radiotherapy alone in the treatment of esophageal cancer. In an RTOG randomized trial involving 129 patients with esophageal cancer, irradiation (50 Gy) with concurrent cisplatin and 5-FU provided a significant survival advantage (27% vs 0% at 5 years) and improved local control over radiation therapy alone (64 Gy). Median survival also was significantly better in the combined-therapy arm than in the irradiation arm (14.1 vs 9.3 months) (Cooper et al., 1999).

A Cochrane review confirmed the superiority of chemoradiotherapy versus radiotherapy in fit patients (Rebecca, 2003).

3.7. Radiotherapy

Radiotherapy is one of the main, effective and relatively safe treatment modalities for cancer esophagus. It could be used for early stage and advanced diseases and as locally palliative treatment for metastatic disease.

The Radiation Therapy Oncology Group (RTOG) 85-01 trial was a randomized controlled comparison of definitive radiotherapy alone (64 Gy), and definitive concurrent chemoradiation (50 Gy delivered concurrently with 5-fluorouracil [5-FU] and cisplatin). A statistically significant benefit was noted for overall survival among patients receiving concurrent chemoradiation (Cooper et al., 1999).The Intergroup trial 0123 subsequently randomized 231 patients to receive definitive chemoradiation with 50 Gy delivered concurrently with 5-FU and cisplatin vs. 64 Gy delivered concurrently with the same chemotherapeutic regimen. No significant differences were noted in median or overall survival or locoregional control (al-Sarraf et al., 1997).Given these findings, the current standard of care for inoperable esophageal cancer is concurrent chemoradiation with 50 Gy radiotherapy.

3.7.1. 3D Conformal Radiotherapy

Three-dimensional conformal radiation therapy (3-DCRT) has been demonstrated to improve dose distribution, thereby allowing significant increase of target dose and decrease of lung and heart doses.

Target volume delineation:

It is based on the International Commission on Radiation Units and Measurements (ICRU)-50 definitions of gross tumour volume (GTV), clinical target volume (CTV), and planning target volume (PTV). To cover both submucosal tumour spread and lymphatics along the oesophagus, enlarged longitudinal safety margins have been validated by clinical and pathological reviews (Hosch et al., 2001 & Gao et al., 2007).

GTV (gross tumor volume) is tumor extension visible in imaging, including primary tumor and enlarged lymph nodes. The commonly used imaging methods include endoscope, esophagogram, CT, MRI; PET-CT. Complementary effect exists between each imaging examination method, and could significantly improve the accuracy and sensitivity when judging the gross tumor volume. Many studies recommend PET-CT for planning simulation (Konshi et al., 2005 & Moureau-Zabotto et al., 2005). Leong et al., (2006) enrolled 21 esophageal carcinoma patients in a prospective trial to determine effects of PET-CT on delineation of tumor volume for radiation therapy planning. PET-CT detected disease in eight patients that was not detected by CT scan: four of these patients were found to have metastatic disease and four had regional nodal disease. In 16 of 21 patients who proceeded to the radiotherapy planning phase of the trial, 69%had PET-CT–positive disease that would have been excluded if CT alone had been used for radiation planning. In cases where an endoscope is unable to pass through a stenosed oesophagus to visualize the lower boundary of the tumour, PET may be the only way to estimate the lower border of the tumour. PET has a significant impact on GTV and PTV in oesophageal cancer, often helping to avoid geographic miss by identifying unsuspected lymph node involvement.

CTV (clinical target volume) refers to the range of subclinical lesions. The microscopic infiltration ranges were < 3 cm superior and inferior along the vertical axis of esophagus in 94% of the patients with esophageal carcinoma as reported by Gao et al. (2007) who concluded that a 50 mm CTV would be necessary to cover distal microscopic spread in 94% of adenocarcinomas of the gastroesophageal junction. A 30 mm CTV would be adequate to cover microscopic disease spread in 94% of squamous cell carcinomas and for coverage of proximal microscopic spread for adenocarcinomas of the gastroesophageal junction.

Clinical target volume node refers to the lymphatic drainage districts of esophageal carcinoma. There is no high grade evidence identifying the range of lymphatic drainage districts in prophylactic radiation for esophageal carcinoma. The final CTV may be larger since for cervical primaries; the supraclavicular nodes need to be included; and for distal primaries, the celiac nodes need to be included.

Planning Target Volume (PTV) will provide margin around the CTV to compensate for variations in treatment set-up, and organ motion will be included in the treatment fields.

A volumetric treatment planning CT study is required to define GTV and PTV. The local regional nodes will be included in the clinical target volume (CTV). Each patient will be positioned in an individualized immobilization device in the treatment position on a flat table. Contiguous CT slices, 3-5 mm thickness of the regions harboring gross tumor and grossly enlarged nodes and 8-10 mm thickness of the remaining regions, are to be obtained starting from the level of the cricoid cartilage and extending inferiorly through the liver. The GTV and PTV and normal organs are outlined on all appropriate CT slices and displayed using beam's eye view. Normal tissues to be contoured include lungs, kidneys, skin, heart, spinal cord, esophagus, and liver. A measurement scale for the CT image shall be included. Barium swallow during the planning CT is optional provided a diagnostic chest CT was done with contrast to delineate the outline of the esophagus (RTOG 0436).

Variability in treatment setup, breathing, or motion during treatment:

Lorchel et al. (2006) and Yaremko et al. (2008) reported that the movement range of esophagus in all directions was 0.5 cm in upper part, 0.6 - 0.7cm in middle part, and 0.8 - 0.9 cm in lower part. A margin around the CTV will define the PTV. The PTV volume must include a minimum of 1 cm and a maximum of 2 cm around the CTV. Once again, the final PTV may be larger, since the supraclavicular nodes need to be included in the treatment fields for cervical primaries and the celiac nodes need to be included in the treatment fields for distal primaries.

The ICRU reference point is to be located in the central part of PTV. Typically, this point should be located on the beam axis or at the intersection of the beam axis (isocenter).

Radiotherapy doses:

External beam radiation therapy to a total dose of 50.4 Gy at 1.8 Gy/fraction, in combination with concurrent cisplatin + 5-FU chemotherapy is currently the standard regimen for definative treatment. The Intergroup 0123 trial randomly assigned 236 patients with locally advanced esophageal cancer (T1-4, N0/1) to radiation to a total dose of 50.4 Gy or 64.8 Gy at 1.8 Gy/ fraction. Concurrent chemotherapy (cisplatin +5-FU) was used in both groups. The results revealed no differences in locoregional failure rates (56% versus 52%) and 2-year overall survival rates (31% versus 40%), as well as in median survival (13 months versus 18 months) (Minsky et al., 2002). analysis of RTOG 94-05 did show that a dose of 64.8 Gy was not superior to 50.4 Gy (Wither and Peters, 1980). Whba et.al (2012) reported overall response rate 54.8% and median overall survival 20 months on using chemoradiotherapy with capecitabine, oxaliplatin and radiotherapy dose 45Gy. RTOG 0436 trial recommend a total dose of 50.4 Gy (1.8 Gy/Fx/day), the prescription dose will be specified at the ICRU-50 reference point; this point will usually be the isocenter (intersection of the beams). The isodose curve representing 93% of the prescription dose must encompass the entire planning target volume (PTV). The daily prescription dose will be 1.94 Gy at the International Commission on Radiation Units and Measurement (ICRU) reference point. 1.8 Gy (which corresponds to the 93% isodose curve) is to be delivered to the periphery of the PTV.

In the trial by Bosset and coworkers (1997) the fractionation consisted of two 1-week courses of 3.7 Gy /5 fractions, the field included the tumor with 5-cm superior and inferior margins and 2 cm radial margins. The celiac axis was not included. Walsh and coinvestigators (2006) used a dose of 40 Gy in 2.67-Gy fractions. The Cancer and Leukemia Group B 9781 trial (Krasna et al., 2006) treated to 50.4 Gy. Radial margins were 2 cm beyond the esophagus; superior and inferior field borders were 5 cm above and below the gross tumor, including the supraclavicular nodes for proximal tumors and the celiac axis for distal tumors.

The fractination schemes, with 50.4 Gy commonly used in the United States and lower doses with larger fraction sizes more common in Europe (Hong et al., 2007).

RTOG9504 established 50.4 Gy as the standard dose of radiation to be administered concurrently with chemotherapy (Minsky et al., 2002).

Field arrangement:

The preferable method is a 3-field technique (2 anterior obliques and a posterior field). In most cases, this is not possible; therefore, it is acceptable to initially treat AP/PA to approximately 39.6 Gy, then switch to obliques to exclude the spinal cord. The supraclavicular field, which is excluded from the obliques, can be supplemented with electrons to bring the total dose up to 50.4 Gy (RTOG 0436). A common approach is anterior: posterior (APPA) fields for the first course, and a 3-field approach consisting of an AP and 2 posterior obliques, or opposed obliques, for the cone-down volume. The advantages of this approach include limiting the lung dose during the AP-PA portion of treatment and then limiting the spinal cord dose by replacing the PA field with off-cord obliques. A disadvantage is the significant cardiac volume often included in treatment filed (Hong et al., 2007).

Dose constraint:

The dose limitation for critical structures includes, the spinal cord dose limited to 45 Gy, 60% of the liver should not exceed 30 Gy, at least two thirds of one kidney should not exceed more than 20 Gy, and one third of the heart should receive less than 50 Gy (Wither and Peters, 1980). The mean lung dose should not exceed 20 Gy, and specific limits have been recommended for volumes that receive 10, 20, and 30 Gy, respectively (V10, V20, V30) (Hong et al., 2007).

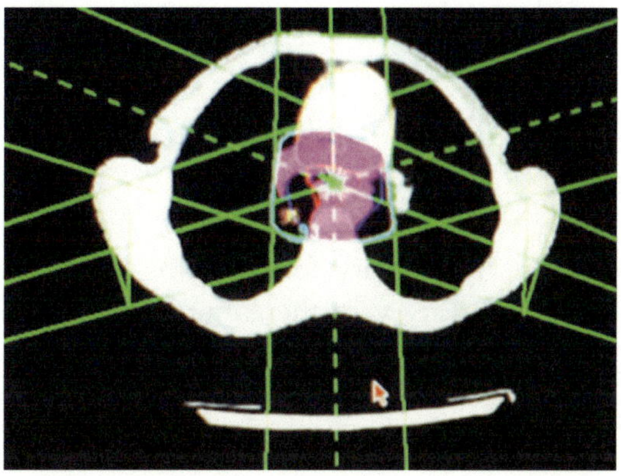

a

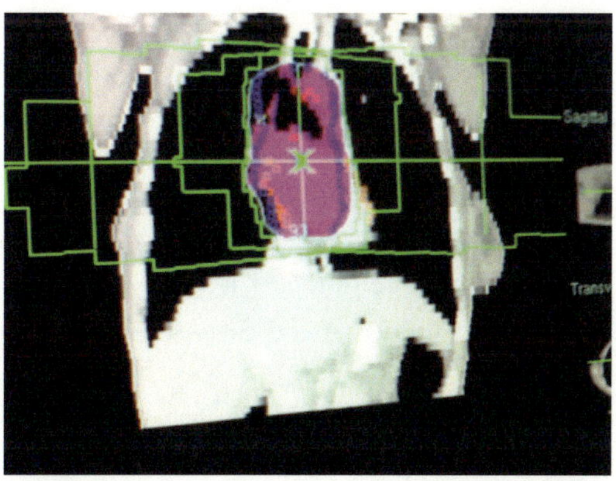

b

Figure 1. a, b Dose distribution by using an AP field and two oblique fields.

3.7.2. Intensity Modulated Radiation Therapy (IMRT)

Most data regarding IMRT for esophageal malignancies has been limited to dosimetric analyses.

Wang et al (2006) reported outcomes of seven patients with locally advanced upper or cervical esophageal cancer treated definitively with concurrent chemoradiation with a total radiation dose of 59.4–66 Gy five- to nine-beam IMRT were used to deliver a total dose of 59.4-66 Gy (median: 64.8 Gy) to the primary tumor. After median fellow up period 15 months all 6 evaluable patients achieved complete response. Of them, 2 developed local recurrences and 2 had distant metastases, 3 survived with no evidence of disease. After treatment, 2 patients developed esophageal stricture requiring frequent dilation and 1 patient developed tracheal-esophageal fistula.

Another study conducted by Fu et al (2004) comparing IMRT and 3D Conformal radiotherapy, The IMRT plans were superior in that they reduced the percent of total lung volume exceeding 20 Gy (V20) or 30 Gy (V30) while generating more conformal and homogeneous target coverage. Heterogeneity and conformality indices were improved with IMRT. No significant reductions were noted in heart, spinal cord, liver and total body integral dose (Chandra et al., 2005).

3.7.3. Follow up after radiotherapy

One month after radiotherapy then every four months in the first year, once every six months in the second year, once every year thereafter to at least five years. If the patients have symptoms they should be followed up according to clinical requirement. Evaluations during these follow-up visits included blood routine, biochemistry test, upper gastroenterography and/or esophagoscopy, chest- X ray films/chest computed tomography (CT).

3.8. Targeted therapy

Bonner et al. (2006) conducted a phase III trial in locally advanced head and neck cancer patients documented the benefit of combination of cetuximab and radiation and reported an improvement in both local control and overall survival. The Brown University Oncology Group and the University of Maryl and Greenebaum Cancer Center (Suntharalingam et al., 2006) have piloted the addition of cetuximab with a concurrent chemoradiation (weekly carboplatin, paclitaxel, and 50.4 Gy), this phase II trial reported complete response rate of 65% in patients presenting with locally advanced disease with no grade 4 toxicities and 20% grade 3 esophagitis.

4. Palliative treatment of locally advanced esophageal cancer

The majority of patients presented with locally advanced or metastatic disease which is difficult to control. Considering this fact, it is important to offer treatment providing adequate and rapid palliation of symptoms especially the obstructive symptoms which reflect on the quality of life.

Several management options have been developed to palliate malignant dysphagia. These include endoluminal stenting or surgery and external beam radiation, brachytherapy, chemotherapy, chemoradiotherapy, laser treatment, photodynamic therapy or ablation using injection of alcohol or chemotherapeutic agents (Weigel et al., 2002; Allum et al., (2002) & Bown (1991).

4.1. Surgical bypass

Surgical bypass advocated as a palliative modality, particularly when unresectability is only discovered at an attempt at curative resection, on the basis that this offers better restoration of swallowing than many palliative modalities (Segslin et al., 1989). However, operative mortality is 40% or more, whichever bypass procedure is used (Segalin et al., 1989 & Whouley et al., 2002).

4.2 Stenting

Since 1990, several case series, retrospective reviews, and prospective studies including more than 2,000 patients have shown that Self-expandable metal stents (SEMS) are effective in relieving dysphagia and improving dysphagia scores, with immediate success rates between 96% and 100% (Dua, 2007).

Madhusudan et al. (2009) reported significant improvements in all QOL parameters after stent placement in patients with advanced inoperable esophageal cancer. This improvement was maintained until 8 weeks.

Another prospective study by Maroju et al. (2006) reported similar improvement in QOL following stenting.

Contraindications to stent use include particularly exophytic tumors, proximal tumors due to pharyngeal irritation caused by prostheses sited too proximally. Complications include tumor ingrowth (predominantly with uncovered stents), tumor overgrowth at the stent margins, and stent migration (particularly with covered stents and lesions close the esophagogastric junction) (Hills et al., 1998 & Tytgat et al., 1986).

4.3 External beam radiotherapy

External beam radiotherapy (EBRT) is known to provide durable and effective relief of dysphagia. However, there is a time lag before symptomatic relief occurs, and up to 6 weeks are required for maximum benefit (Bown, 1991).

Most studies have used radiotherapy in a dose range of 40–60 Gy. However, a higher dose of radiotherapy does not add to the therapeutic value, and may increase the loco-regional toxicity (Minsky et al., 2002).

4.4. Stenting and external beam radiotherapy:

Although SEMS are easy to place and the beneficial effects are immediate, recurrent dysphagia has been observed in many patients during a follow-up period of 4–10 weeks (Homann et al., 2008). Radiotherapy, on the other hand, provides long-term relief of dysphagia (Bown, 1991). Zhong et al. (2003) and Han et al. (2004) have investigated the effect of combined stenting and radiotherapy on survival of patients with advanced esophageal cancer and reported superior results with regard to both relief of dysphagia and survival for stenting followed by radiotherapy in those patients.

Eldeeb and El-Hadaad (2012) conducted a prospective study on 91 locally advanced esophageal cancer patients, they reported median overall survival (OAS) 169 days in radiotherapy group (the radiation doses ranged from 20Gy/5fractions to 30Gy/10 fractions), 119 days in stenting group and 237 days in combined radiotherapy- stenting group, the difference between radiotherapy group and combined radiotherapy- stenting group was significant.

4.5. Thermal ablative therapy

Laser photocoagulation has been the most studied modality; two studies have demonstrated better palliation using laser photocoagulation, its disadvantages include the necessity to repeat treatment at approximately 6-week intervals (Carter et al., 1992) & Loizou et al., 1991). Laser treatment is best reserved for tumors least amenable to stent placement

4.6. Brachytherapy

The use of intracavitary irradiation (brachytherapy) in doses of 1,800 cGy in the palliation of esophageal cancer has encouraging results (Sur M et al., 1996 & Sur RK et al., 2002). Endoluminal approach, high dose-rate brachytherapy (HDRBT) alone may offer sustained symptomatic relief. An established regimen is two fractions of 8 Gy, each prescribed at 1.0 cm, which has been tested in IAEA-randomized trial; it resulted in a median survival of 237 days and the incidences of strictures (11%) and fistulae (10%) (Sur RK et al., 2002).

The combination of high dose-rate brachytherapy (HDRBT) and External Beam Radiation Therapy (EBRT) is superior to HDRBT alone for the palliation of oesophageal cancer. Addition of EBRT to HDRBT improved dysphagia-relief experience (DRE). The average benefit was an absolute +18% improvement in DRE which was sustained between 50 to 350 days of follow-up. The overall improvement in mean dysphagia score was -0.44. While HDRBT alone produced, on average, a relatively stable dysphagia score, the addition of EBRT led to a further reduction in the score compared with that from HDRBT alone (Rosenblatt et al., 2010).

5. Conclusion and future recommendations

The treatment of locally advanced esophageal cancer is a multidisciplinary approach; Radiochemotherapy is the standard of care; combined chemotherapy and radiation therapy is the definitive treatment of choice for unresectable or medically inoperable locally advanced esophageal

Cancer, neoadjuvant chemoradiation therapy improves OAS survival.

In future, it will be crucial to improve chemotherapy regimens, radiation delivery, and surgical techniques to reduce morbidity and mortality and to increase cure rates. All esophageal cancer patients should be managed at a center where a multidisciplinary setting is well established (Kaifi et al., 2011).

To improve treatments outcomes, utilization of modern radiation therapy technology including respiratory gating, image guidance and IMRT allow high precise localization, greater dose of radiation to target and decreasing normal tissues toxicity. Incorporation of newer chemotherapeutic agents into chemoradiation regimens, also using targeted therapies in combination with chemoradiotherpy may be having promising results.

Author details

Hend Ahmed El-Hadaad and Hanan Ahmed Wahba
Clinical Oncology and Nuclear Medicine, Faculty of Medicine, Mansoura University, Mansoura, Egypt

6. References

Ajani JA, Walsh G, Komaki R et al (2004) Preoperative induction of CPT-111 and cisplatin chemotherapy followed by chemoradiotherapy in patients with locoregional carcinoma of the esophagus or gastroesophageal junction. Cancer 100:2347–354.

Allum WH, Griffin SM, Watson A et al (2002) Guidelines for the management of oesophageal and gastric cancer. Gut; 50 Suppl 5: v1-v23.

al-Sarraf M, Martz K, Herskovic A et al (1997) Progress report of combined chemoradiotherapy versus radiotherapy alone in patients with esophageal cancer: an intergroup study. J Clin Oncol. 15:277–284.

Arnott SJ, Duncan W, Gignoux M et al (2005) Preoperative radiotherapy for esophageal carcinoma. Cochrane Database Syst Rev. CD001799.

Arnott SJ, Duncan W, Kerr GR et al (1992) Low dose preoperative radiotherapy for carcinoma of the oesophagus: results of a randomized clinical trial. Radiother Oncol. 24:108-113.

Berger AC, Farma J, Scott WJ et al (2005) Complete response to neoadjuvant chemoradiotherapy in esophageal carcinoma is associated with significantly improved survival. J Clin Oncol. 23:4330-4337.

Boige V, Pignon J, Saint-Aubert B et al (2007) Final results of a randomized trial comparing preoperative 5-fluorouracil (F)/cisplatin (P) to surgery alone in adenocarcinoma of stomach and lower esophagus (ASLE): FNLCC ACCORD07-FFCD 9703 trial. J Clin Oncol. 25(18S):Abstract 4510.

Bonner JA, Harari PM, Giralt J et al (2006) Radiotherapy plus cetuximab for squamous-cell carcinoma of the head and neck. N Engl J Med 354:567- 578.

Bosset JF, Gignoux M, Triboulet JP et al (1997) Chemoradiotherapy followed by surgery compared with surgery alone in squamous-cell cancer of the esophagus. N Engl J Med 337:161-167.

Bown SG. (1991): Palliation of malignant dysphagia: surgery, radiotherapy, laser, intubation alone or in combination? Gut; 32: 841-844.

Brücher BL, Becker K, Lordick F et al (2006) The clinical impact of histopathologic response assessment by residual tumor cell quantification in oesophegeal squamous cell carcinoma. Cancer. 106:2119–27.

Carter R, Smith JS, Anderson JR (1992) Laser recanulization versus endoscopic intubation in the palliation of malignant dysphagia: a randomised prospective study. Br J Surg 79:1167–1170.

Caspar LE, Winter K, Kocha WI et al (2000) A phase I/II study of external beam radiation, brachytherapy, and concurrent chemotherapy for patients with localized carcinoma of the esophagus (Radiation Therapy Oncology Group Study 9207): final report. Cancer 88:988-95.

Chandra A, Guerrero TM, Liu HH et al (2005) Feasibility of using intensity-modulated radiotherapy to improve lung sparing in treatment planning for distal esophageal cancer. Radiother Oncol.77:247–253.

Choi JY, Lee KH, Shim YM et al (2000) Improved detection of individual nodal involvement in squamous cell carcinoma of the esophagus by FDG-PET. J Nucl Med 41:808–815.

Coia LR, Engstrom PF, Paul AR et al (1991) Long-term results of infusional 5-FU, mitomycin-C and radiation as primary management of esophageal carcinoma. Int J Radiat Oncol Biol Phys. 20:29-36.

Cooper JS, Guo, Herskovic A et al (1999) Chemoradiotherapy of locally advanced esophageal cancer long term follow –up of a prospective randomized trial (RTOG 85-01). JAMA 281:1623-1627.

Cunningham D, Allum WH, Stenning SP et al (2006) Perioperative chemotherapy versus surgery alone for resectable gastroesophageal cancer. N Engl J Med. 355:11-20.

Downey RJ, Akhurst T, Ilson D et al (2003) Whole body 18FDG-PET and the response of esophageal cancer to induction therapy: results of a prospective trial. J Clin Oncol. 21:428-432.

Dua KS (2007) Stents for palliating malignant dysphagia and fistula: is the paradigm shifting? Gastrointest Endosc. 65(1):77–81.

Eisterer W, Kendler D, De Vries A et al (2011) Triple induction chemotherapy and chemoradiotherapy for locally advanced esophageal cancer. A phase II study Anticancer Research 12:4407-4412.

Eldeeb H, El-Hadaad HA (2012) Radiotherapy Versus Stenting In treating malignant dysphagia. J Gastrointest Oncol. 3(4):322-5. DOI: 10.3978/j.issn.2078-6891.2012. 011.

Falk J, Carstens H, Lundell L et al (2007) Incidence of carcinoma of the oesophegus and gastric Cardia-Changes over time and geographical differences. Acta Oncol. 46:1070–4.

Flamen P, Van Cutsem E, Lerut A et al (2002) Positron emission tomography for assessment of the response to induction radiochemotherapy in locally advanced oesophageal cancer. Ann Oncol. 13:361-368.

Fu WH, Wang LH, Zhou ZM et al (2004) Comparison of conformal and intensity-modulated techniques for simultaneous integrated boost radiotherapy of upper esophageal carcinoma. World J Gastroenterol. 10:1098–1102.

Gao XS, Qiao X, Wu F et al (2007) Pathological analysis of clinical target volume margin for radiotherapy in patients with esophageal and gastroesophageal junction carcinoma. Int J Radiat Oncol Biol Phys 67:389–96.

Han YT, Peng L, Fang Q et al (2004) Value of radiotherapy and chemotherapy after SEMS implantation operation in patients with malignant esophageal stricture. Ai Zheng 23(6): 682–4.

Heath EI, Burtness BA, Heitmiller RF et al (2000) Phase II evaluation of preoperative chemoradiation and postoperative adjuvant chemotherapy for squamous cell and adenocarcinoma of the esophagus. J Clin Oncol.18:868-876.

Herskovic A, Martz K, al-Sarraf M et al (1992) Combined chemotherapy and radiotherapy compared with radiotherapy alone in patients with cancer of the esophagus. N Engl J Med 326:1593–1598.

Hills KS, Chopra KB, Pal A et al (1998) Self-expanding metal oesophageal endoprostheses covered and uncovered: a review of 30 cases. Eur J Gastroenterol Hepatol 5:367–370.

Homann N, Noftz MR, Klingenberg-Noftz RD et al (2008) Delayed complications after placement of self-expanding stents in malignant esophageal obstruction: treatment strategies and survival rate. Dig Dis Sci. 53(2):334–40.

Hong TS, Crowley EM, Killoran J et al (2007) Considerations in treatment planning for esophageal cancer. Semin Radiat Oncol 17:53–61.

Hosch SB, Stoecklein NH, Pichlmeier U et al (2001) Esophageal cancer: the mode of lymphatic tumor cell spread and its prognostic significance. J Clin Oncol 19:1970–5.

Kaifi JT, Gusani NJ, Jiang Y et al [2011] Multidisciplinary Management of Early and Locally Advanced Esophageal Cancer.J Clin Gastroenterol 45:391–399.

Kaklamanos IG, Walker GR, Ferry K et al (2003) Neoadjuvant treatment for resectable cancer of the esophagus and the gastroesophageal junction: A meta-analysis of randomized clinical trials. Ann Surg Oncol 10:754–761.

Kato H: Kuwano H, Nakajima M et al (2002) Usefulness of positron emission tomography for assessing the response of neoadjuvant chemo-radiotherapy in patients with esophageal cancer. Am J Surg 184:279–283.

Khushalani NI, Leichman CG, Proulx G et al (2002) Oxaliplatin in combination with protracted infusion fluorouracil and radiation: report of a clinical trial for patients with oesophegeal cancer.J Clin Oncol. 20:2844–50.

Konski A, Doss M, Milestone B et al (2005) The integration of 18-fluoro-deoxy-glucose positron emission tomography and endoscopic ultrasound in the treatment-planning process for esophageal carcinoma. Int J Radiat Oncol Biol Phys. 61:1123–1128.

Krasna M, Tepper JE, Niedzwiecki D et al (2006) Trimodality therapy is superior to surgery alone in esophageal cancer: Results of CALGB 9781. Paper presented at: ASCO 2006 Gastrointestinal Cancers Symposium, San Francisco, California, January 26-28.

Leong T, Everitt C, Yuen K et al (2006) A prospective study to evaluate the impact of FDG-PET on CT based radiotherapy treatment planning for oesophageal cancer. Radiother Oncol. 78:254– 261.

Lerut T, Coosemans W, Decker G et al (2001) Cancer of the esophagus and gastroesophegeal junction: potentially curative therapies. Surg Oncol. 10:113–22.

Lerut T, Flamen P, Ectors N et al (2000) Histopathological validation of lymph node staging with FDG-PET in cancer of the esophagus and gastro-esophageal junction. Ann Surg 232:743–752.

Loizou LA, Grigg D, Atkinson M et al (1991) A prospective comparison of laser therapy and intubation in endoscopic palliation of malignant dysphagia. Gastroenterology 100:1303–1310.

Lorchel F, Dumas JL, Noel A et al (2006) Esophageal cancer: determination of internal target volume for conformal radiotherapy. J Radiother Oncol, 80(3):327-332.

Lordick F, Ott K, Krause BJ et al (2007) PET to assess early metabolic response and to guide treatment of adenocarcinoma of the oesophagogastric junction: the MUNICON phase II trial. Lancet Oncol. 8:797-805.

Madhusudan C, Saluja SS, Pal S et al (2009) Palliative stenting for relief of dysphagia in patients with inoperable esophageal cancer: impact on quality of life. Dis Esophagus. 22(4):331–6.

Makary MA, Kiernan PD, Sheridan MJ et al (2003) Multimodality treatment for esophageal cancer: the role of surgery and neoadjuvant therapy. Am Surg. 69:693-700; discussion 700-692.

Maroju NK, Anbalagan P, Kate V et al (2006) Improvement in dysphagia and quality of life with self-expanding metallic stents in malignant esophageal strictures. Indian J Gastroenterol. 25(2):62–5.

Medical Research Council Oesophageal Cancer Working Group (2002) Surgical resection with or without preoperative chemotherapy in oesophageal cancer: a randomised controlled trial.Lancet. 359:1727-1733.

Meluch AA, Greco FA, Gray JR et al (2003) Preoperative therapy with concurrent paclitaxel/carboplatin /infusional 5- FU and radiation therapy in locoregional esophageal cancer: final results of a Minnie Pearl Cancer Research Network phase II trial. Cancer J.9:251-260.

Minsky BD, Pajak TF, Ginsberg RJ et al (2002) INT0123 (Radiation Therapy Oncology Group 94-05) phase III trial of combined-modality therapy for esophageal cancer: High-dose versus standard-dose radiation therapy. J Clin Oncol. 20:1167-1174.

Moureau-Zabotto L, Touboul E, Lerouge D et al (2005) Impact of CT and 18F-deoxyglucose positron emission tomography image fusion for conformal radiotherapy in esophageal carcinoma. Int J Radiat Oncol Biol Phys. 63:340–345.

National Comprehensive Cancer Network guidelines (2009) Clinical practice guidelines in oncology, Esophageal cancer. Available at: http://www.nccn.org/ Accessed 07 January 2009.

Nguyen P, Feng JC, Chang KJ (1999) Endoscopic ultrasound (EUS) and EUSguided fine-needle aspiration (FNA) of liver lesions. Gastrointest Endosc 50:357–361.

Nygaard K, Hagen S, Hansen HS et al (1992) Pre-operative radiotherapy prolongs survival in operable esophageal carcinoma: a randomized, multicenter study of pre-operative radiotherapy and chemotherapy.The second Scandinavian trial in esophageal cancer. World J Surg.16:1104-1109; discussion 1110.

Oh CE, Antes K, Darby M et al (1999) Comparison of 2D conventional, 3D conformal, and intensity-modulated treatment planning techniques for patients with prostate cancer with regard target-dose homogeneity and dose critical, uninvolved structures. Med Dosim 24:255–263.

Ott K, Weber WA, Lordick F et al (2006) Metabolic imaging predicts response, survival, and recurrence in adenocarcinomas of the esophagogastric junction. J Clin Oncol. 24:4692-4698.

Polee M, Tilanus HW, Eskens FA et al (2003) Phase II study of neo-adjuvant chemotherapy with paclitaxel and cisplatin given every 2 weeks for patients with a resectable squamous cell carcinoma of the esophagus. Ann Oncol 14:1253–1257.

Rebecca WO, Richard MA (2003) Combined chemotherapy and radiotherapy (without surgery) compared with radiotherapy alone in localized carcinoma of the esophagus. Cochrane Database Sys Rev (1):CD002092.

Rosenblatt E, Jones G, Sur RK et al (2010) Adding external beam to intra-luminal brachytherapy improves palliation in obstructive squamous cell oesophageal cancer: A prospective multi-centre randomized trial of the International Atomic Energy Agency. Radiotherapy and Oncology 97: 488–494.

Segalin A, Little AG, Ruol A et al (1989) Surgical and endoscopic palliation of esophageal carcinoma. Ann Thorac Surg 48:267–271.

Siegel R, Ward E, Brawley O, Jemal A (2011) Cancer statistics, 2011: The impact of eliminating socioeconomic and racial disparities on premature cancer deaths. CA Cancer J Clin 61:212.

Stahl M, Stuschke M, Lehmann N et al (2005) Chemoradiation with and without surgery in patients with locally advanced squamous cell carcinoma of the esophagus. J Clin Oncol. 23:2310-2317.

Suntharalingam M, Dipetrillo T, Ackerman P et al (2006) Cetuximab, paclitaxel, carboplatin, and radiation for esophageal and gastric cancer. Proc Am Soc Clin Oncol 24:185s, (abstr 4029).

Sur M, Sur RK, Cooper K et al (1996) Morphologic alterations in esophageal squamous cell carcinoma after preoperative high dose rate intraluminal brachytherapy. Cancer 77:2200–2205.

Sur RK, Levin CV, Rad FF et al (2002) Prospective randomised trial of HDR brachytherapy as a sole modality in palliation of advanced esophageal carcinoma: an International Atomic Energy Agency study. Int J Radiat Oncol Biol Phys 53:127–133.

Swisher SG, Hofstetter W, Wu TT et al (2005) Proposed revision of the oesophegeal cancer staging system to accommodate pathologic response (PP) following pre-operative chemoradiation. Ann Surg. 241:810–20.

Tepper J, Krasna MJ, Niedzwiecki D, et al (2008) Phase III trial of trimodality therapy with cisplatin, fluorouracil, radiotherapy and surgery compared with surgery alone for oesophegeal cancer. CALGB 9781. J Clin Oncol. 26(7):1086–92.

Tytgat GNJ, den Hartog Jager FC et al (1986) Endoscopic prosthesis for advanced esophageal cancer. Endoscopy 18:32–39.

Valerdi JJ, Tejedor M, Illarramendi JJ et al (1993) Neoadjuvant chemotherapy and radiotherapy in locally advanced esophagus carcinoma: long-term results. Int J Radiat Oncol Biol Phys. 27:843-847.

Wahba HA, El-Hadaad HA, Abd-Ellatif EA (2012) Neoadjuvant concurrent chemoradiotherapy with capecitabine and oxaliplatin in patients with locally advanced esophageal cancer. Med Oncol. 29:1693-1698.

Wakelin SJ, Deans C, Crofts TJ et al (2000) A comparison of computerised tomography, laparoscopic ultrasound and endoscopic ultrasound in the preoperative staging of oesophago-gastric carcinoma. Eur J Radiol 41:161–167.

Walsh TN, Noonan N, Hollywood D et al (1996) A comparison of multimodal therapy and surgery for esophageal adenocarcinoma. N Engl J Med 335:462-467.

Wang SL, Liao Z, Liu H et al (2006) Intensity-modulated radiation therapy with concurrent chemotherapy for locally advanced cervical and upper thoracic esophageal cancer. World J Gastroenterol. 12:5501–5508.

Weber WA, Ott K, Becker K et al (2001) Prediction of response to preoperative chemotherapy in adenocarcinomas of the esophagogastric junction by metabolic imaging. J Clin Oncol. 19:3058-3065.

Weigel TL, Frumiento C, Gaumintz E (2002) Endoluminal palliation for dysphagia secondary to esophageal carcinoma. Surg Clin North Am; 82: 747-761.

Whouley BP, Law S: Murthy SC et al (2002) The Kirschner operation in unresectable esophageal cancer. Arch Surg 137:1z28–1232.

Withers HR, Peters LJ (1980) Basic principles of radiotherapy: basic clinical parameters. In: Fletcher GA, editor. Textbook of radiotherapy. Philadelphia: Lea & Febiger p. 180.

Yaremko BP, Guerrero TM, McAleer MF et al (2008) Determination of respiratory motion for distal esophagus cancer using four-dimensional computed tomography [J]. Int J Radiat Oncol Biol Phys, 70(1):145-153.

Zhong J, Wu Y, Xu Z et al (2003) Treatment of medium and late stage esophageal carcinoma with combined endoscopic metal stenting and radiotherapy. Chin Med J. 116(1): 24–8.

Chapter 9

Reduction Mammaplasty and Intra-Operative Radiotherapy (IORT) in Conservative Surgery

Simonetta Franchelli, Paolo Meszaros, Michela Massa, Marina Guenzi, Renzo Corvò, Davide Pertile, Giorgia Timon, Ferdinando Cafiero and Pierluigi Santi

Additional information is available at the end of the chapter

http://dx.doi.org/10.5772/55939

1. Introduction

1.1. Conservative surgery

Conservative surgery (lumpectomy, quadranctectomy) associated with radiation therapy is widely applied worldwide instead of mastectomy in early-stage breast cancer, since clinical trials reported similar survival rate [1,2].

These surgical procedures are the first choice for small primary tumours (diameter on physical examination up to 3 cm) with no palpable nodes (N0); the choice of procedure adopted depending upon the relation between tumour size and breast volume (small tumour in voluminous breast).

Contraindications to a conservative surgery are: serious co-morbidities, multifocal tumour or wide microcalcifications, clinically palpable lymph nodes and all contraindications to radiotherapy.

The conservative surgery may expose to the patient to a higher risk of local recurrence, that could be reduced applying wider excision with microscopic clear margins. Although there are controversies with regard to a safe margin, it is generally accepted that there is lower risk of recurrence with clear margin more than 10 mm, while margins less than 2 mm are considered inadequate [3]. Excision with clear margins is important for local control and consequently for overall survival.

In our experience sentinel node biopsy represents another step of conservative surgery because it is important for tumour staging.

1.2. Radiotherapy and IORT

The standard adjuvant therapy in early breast carcinoma includes whole breast irradiation (WBI) after conservative surgery to minimize the risk of local failure and improve disease-specific survival. The standard schedules for WBI is 1.8- to 2-Gy daily fractions given 5 times a week to a total dose of 45 to 50 Gy over 5 weeks with optional addition of a boost to the primary site of 10 to 16 Gy in 5 to 8 daily fractions over 1 to 1.5 weeks.The irradiation takes almost 6 weeks. Recent randomized trials have shown that use of modest hypo-fractionation for adjuvant WBI in women with early breast cancer can reduce the number of weeks of treatment (3 or 4 depending on the schemes used) to obtain [4].

Another aspect is whether whole mammary gland needs to be irradiated to destroy microscopic tumour foci. Few studies have systematically addressed the extent of foci of premalignant and malignant disease in the breast after lumpectomy [5,6]. It has been shown that around 40% of the cases had no other foci in the breast of pre-malignancy/malignancy, thus 60% of the cases had residual foci. Moreover at a distance of >2cms from the primary carcinoma only 14–16% had invasive tumor foci in the breast.

Nevertheless it has been uniformly demonstrated that local recurrence rate is significantly lowered by adjuvant radiotherapy [2] and that the majority of local recurrences (LRs) occurs in proximity to the tumour bed, while less than 20% of LRs appear "elsewhere" in the breast. [7,8]

On the basis of these considerations, there has been a growing interest in the use of accelerated partial-breast irradiation (APBI) as an alternative to WBI in the last decade. This technique irradiates a limited volume of the mammary gland to a high dose in 1 to 10 fractions to be delivered in 1 to 5 days. The advantages of APBI are: a decreased overall treatment time and decreased radiation dose delivered to uninvolved portions of the breast and adjacent organs.

It is important to recognize that APBI is unlikely to replace WBI for all patients treated with breast-conserving surgery and that the key to long-term success for partial breast irradiation is proper selection to identify low risk patients.

The American Society for Radiation Oncology (ASTRO) and the Groupe Européen de Curiethérapie-European Society for Therapeutic Radiology and Oncology (GEC-ESTRO) [9,10] breast cancer working groups developed a consensus statement regarding patient selection criteria for the use of APBI outside the context of a clinical trial. The recommendations were based on the results of a systematic literature review and were supplemented by the expert opinions of both Task Force members.

The inclusion/exclusion criteria of partial breast irradiation after conservative surgery are shown in tab.1-2.

Similar prognostic factors are applied in both the documents developed by two groups: patients factors (young age and BRCA mutations), pathologic factors (tumor size, lobular carcinoma, positive margins of resections, lymph–vascular space invasion, multicentricity, multifocality, pure ductal carcinoma in situ, extensive intraductal component), nodal factors (axillary dissection not performed or pN1, pN2, pN3) and treatment factors as neoadjuvant chemo-therapy. Analyzing the prognostic factors, different categories of risk for local relapse were drawn to exclude patients at high risk from APBI.

Inclusion criteria*	• Non lobular carcinoma in histological specimen after standard lumpectomy; • Age ≥45 and <85 years • Lump diameter ≤ 2 cm (pT1-T2 according to TNM-UICC 2002); • Safe resection margin ≥5 mm in the surgical specimen; • Patient availability to out-patient follow-up; • Patient availability for following medical examinations such as MNR, mammography, ultrasound; • Agreement to informed consent; • pN0.

Table 1.

Exclusion criteria	• Extensive intraductal component in the histological specimen (> 3 cm of diameter); • Pure ductal carcinoma or lobular carcinoma; • Multifocality; • Neoadjuvant therapy • Serious diseases; • Pregnancy or breast feeding; • Psychiatric illness; • Connective tissue diseases

Table 2.

Several technique may be used for partial breast irradiation: interstitial brachytherapy, brachytherapy using MammoSite device, 3D conformal external radiotherapy (3D CRT) and Intraoperative Radiotherapy (IORT). The above mentioned methods for APBI have different characteristics and they are very difficult to compare [11].

IORT is the only APBI technique that offers surgery and radiotherapy simultaneously with great comfort for patients. It can be implemented using low energy photons (50 kV) provided by an Intra-beam device [12] or using accelerated electrons produced by mobile linear accelerators installed in the operating room (fig.1).

Some centers use beams of electrons produced by linear accelerators located outside the operating room; although this process is feasible, having a dedicated accelerator in the operating room facilitates the procedures and reduces discomfort and complications.

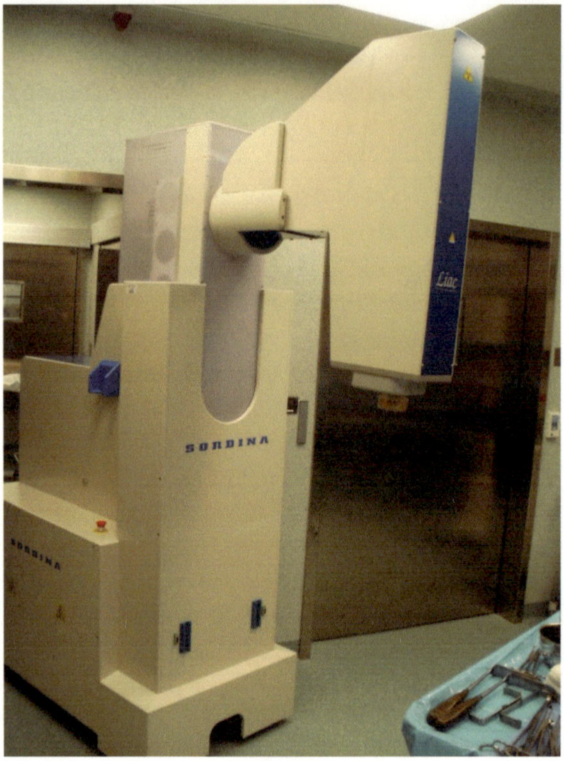

Figure 1. Mobile linear accelerator in our operative room

IORT with electrons is a method for adjuvant radiotherapy developed in 1999 at European Institute of Oncology (IEO) in Milan: Intra-operative Radiotherapy with Electrons (ELIOT) delivers a single dose of radiation equivalent to the total dosage with external fractionated radiotherapy directly to the tumour bed after lumpectomy or quadrantectomy using a mobile linear accelerator located in the operating theatre.

This technique has the advantage to do radiation treatment when breast tissue is having a rich vascular supply; thereby making it more sensitive to the action of the radiation (oxygen effect), immediately after surgery, before tumour cells have a chance to proliferate.

Moreover, the precise application of one single high dose of irradiation directly to the tumour bed with complete skin sparing has the great advantage of shortening radiotherapy time from 6 weeks to one single fraction.

The biological equivalent dose (BED) for IORT needs to be discussed as for a 20–21 Gy dose given in one fraction there are no radiobiological models that can describe what is happening during such high doses. The tool, most commonly used for determining iso-effective doses is the linear–quadratic (LQ) model, and it is unclear if it is applicable for single fractions >10 Gy. Within this limit, the single dose of 21 Gy is equivalent to a fractionated dose of 65 Gy hence, an increased incidence of severe fibrosis can be expected, nevertheless clinical experiences, reported to date, does not support this hypothesis. [13]

Preliminary results are encouraging, but some questions are still unclear, such as the effect of high single doses on late morbidity and aesthetic result. All experts are waiting for the results of prospective randomized ELIOT trial carried on by IEO, comparing the delivery of 21 Gy intraoperatively versus Whole Breast Irradiation (50Gy/25 fraction) plus additional 10 Gy boost irradiation. In the meantime, Veronesi et al have published the results of a retrospective analysis on 1822 patient, treated from January 2000 to December 2008 at the IEO, after a mean follow up of 36.1 months [14].

Forty-two women (2.3%) developed local recurrence, 24 (1.3%) new primary ipsilateral carcinomas and 26 (1.4%) distant metastases as first event. Local side effects were mainly liponecrosis (4.2%) and fibrosis (1.8%). Forty-six women died (2.5%), 28 with breast carcinoma and 18 with other causes. Five- and ten-year survivals were, respectively, 97.4 and 89.7%. Based on these data, ELIOT seems to be promising technology. Moreover, the authors showed that age, tumor size, numbers of positive nodes, molecular subtype, tumour grade and peritumoural invasion are statistically significant predictors of local relapse [14].

A further study has confirmed the importance of patient selection as indicated by the international documents mentioned above. [9,15].

IORT is a method of radiation treatment that requires close collaboration between all components of a multidisciplinary team: oncologic and plastic surgeons, radiation oncologists and medical physicists.

2. Oncoplastic techniques

Several techniques of reduction mammaplasty have been carried out over the years to correct macromastia [16].

In aesthetic field the choice between different mammaplasty techniques depending upon degree of macromastia; patients who require mild resections of 500 gr per breast can be treated by superior pedicle techniques such as Pitanguy reduction mammaplasty with T-shaped inverted scar (Fig.2) or other superior pedicle techniques such as Marchac, Peixoto and Lassus Techniques that have different extension of residual scars [17].

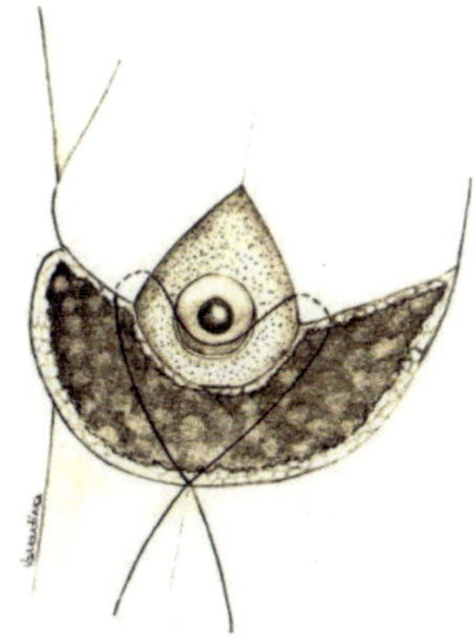

Figure 2. Superior pedicle mammaplasty: gland resection concerns the inferior quadrants

However, patients who require greater resection (1000 gr per breast and more) can be treated more effectively by inferior pedicle techniques or free nipple grafting technique (Fig 3, 4).

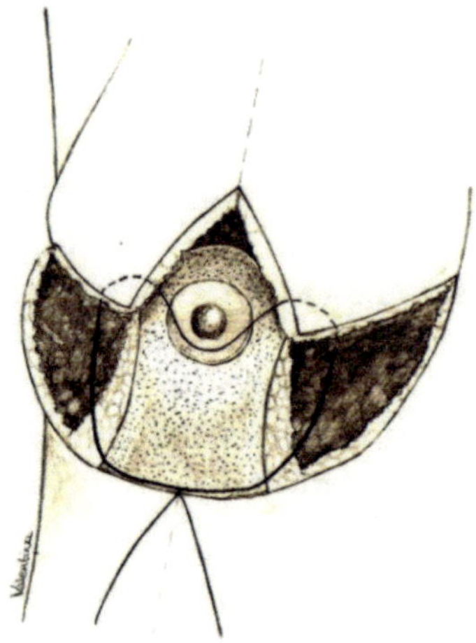

Figure 3. Inferior pedicle mammaplasty: in this case gland resection concerns the superior quadrants. The technique suites better to great reduction.

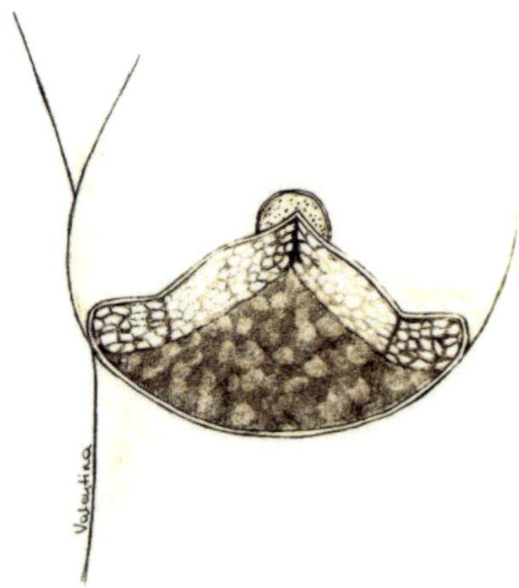

Figure 4. Mammaplasty with free nipple grafting technique: in this case resection concerns the inferior and the central part of the gland

However, in oncologic field the choice between the different options depends primarily on cancer position; for example, the inferior pedicle mammaplasty is chosen, if lump is located in the superior part of the gland; while superior pedicle mammaplasty is applied if lump is located in inferior part of the breast.

This practice minimizes the reshaping and displacement of residual glandular flaps, that may be very important if external radiotherapy is applied.

3. Technique

At our Institute, since 2004 a protocol has been applied to patients with medium/large-sized and ptotic breast (from II to IV degree ptosis) and the patients that needed breast tissue removal of more than 10% of total volume for small breast and more than 20% of total volume for large breast. Different techniques of reduction mammaplasty were carried out in these cases instead of classic conservative surgery (lumpectomy, quadrantectomy).

Oncoplastic techniques have been always applied, regardless of volume/ptosis of breast, if lump is located behind nipple-areola complex.

The introduction of IORT has greatly modified the approach to all features of conservative treatment, both traditional or oncoplastic. Since 2010 we have started to apply oncoplastic techniques of reduction mammaplasty in association with IORT.

The inclusion criteria include criteria of oncoplastic techniques (medium / large breast with/or ptosis III-IV degree) together with the ones of IORT (pts aged >45, T1, T2 < 2.5 cm, intraductal component < 25%, negative lymphatic metastasis, single lump) applied in our Department after approval of Institute Ethical Board.

All the patients underwent oncoplastic procedures and IORT including a multidisciplinary team (oncologic surgeon, plastic surgeon and radiotherapist) during the same surgical time.

A comprehensive preoperative consultation with plastic and oncologic surgeon including a discussion with the patient about her physical peculiarity, psychological status, expectations and choice between unilateral or bilateral procedure, precedes the operation; the choice between unilateral or bilateral mammaplasty is made by the patient and is motivated exclusively by psychological reasons.

A specific informed consent that explains the extent of the undermined tissues and scars and the effect of IORT are discussed with each patient; it explains the different phases of surgery and the possible alternatives (classic conservative surgery), the cutaneous incisions, the extent of the undermined tissues and residual scars.

The possible complications are discussed with the patient, such as: cutaneous and subcutaneous necrosis, seroma, haematoma, numbess and local anesthesia, surgical site infection (SSI), possible asymmetry; moreover some preoperative conditions are evaluated such as smoking that may extend the healing process or may influence the quality of residual scars.

A specific informed consent is administered by radiotherapist that discusses with the patient all different treatment opportunities, the extent of the disease and the agreement to the procedure; this consent includes the explanation of the procedure and the possible complications of IORT, such as mammary oedema, infection, fat necrosis, seroma, late fibrosis.

The markings on the breast are made with patient in standing position and are discussed with the oncologic surgeon clearly showing the position of the lump and the extension of tissues that have to be removed.The patient is consulted with regard to the amount of breast reduction.

Usually the residual scars are T-inverted shaped; although in small reduction mammaplasty, the scar may be only vertical.

4. Surgical and radiotherapy time

The tumour is removed, according to the preoperative drawings, with at least 1 cm of macroscopic margin and submitted to immediate pathological analysis that ensures an adherence to inclusion criteria of IORT (T< 2.5, intraductal component <25%, negative margins > 5 mm). This is followed by a sentinel node biopsy to exclude positive biopsies. If sentinel node biopsy is positive, a complete axillary dissection is performed and IORT is not applied: such a patient is candidate for traditional radiotherapy.

After tumour resection a mobilization of the mammary gland, from the pectoralis fascia and from the skin, is carried out to obtain a good exposure to the radiation beam. A shielding disk (available in various diameter from 4 to 10 cm) is positioned between gland and pectoralis muscle, in line with the collimator to protect thoracic wall, heart and lung (Fig.5)

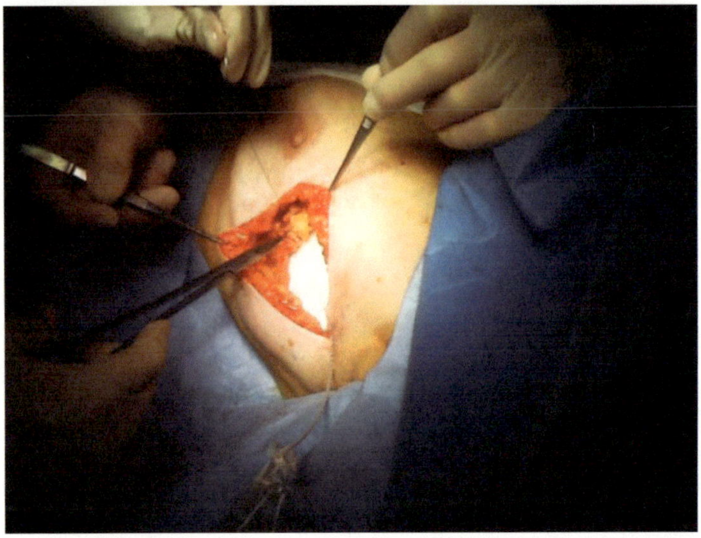

Figure 5. The disk is positioned under the pectoralis fascia and the gland is replaced on it

The disk must be equal or greater in diameter than the collimator which will be used, to realize the best protection of the thoracic wall. The lead disk and collimator diameters are chosen keeping in consideration the ratio of tumour size and breast volume.

The gland is replaced over the disk with temporary and separated stitches so that breast is homogeneous as regard the thickness and receives irradiation in the best way (Fig.6).

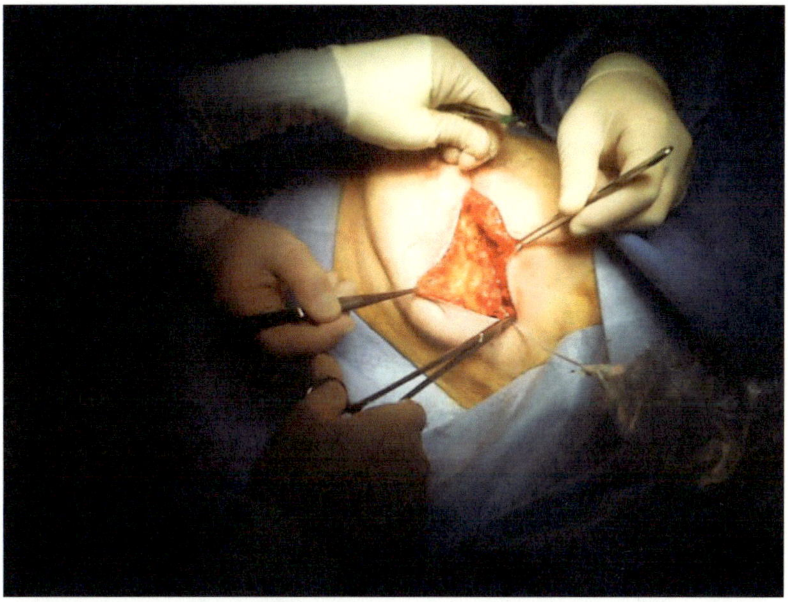

Figure 6. Breast tissues sutured on the disk with temporary stitches

The breast thickness is subsequently measured by graduated needle (perpendicularly inserted through the breast target until the hard surface of the disk can be felt), and the effective electron energy is chosen (Fig. 7).

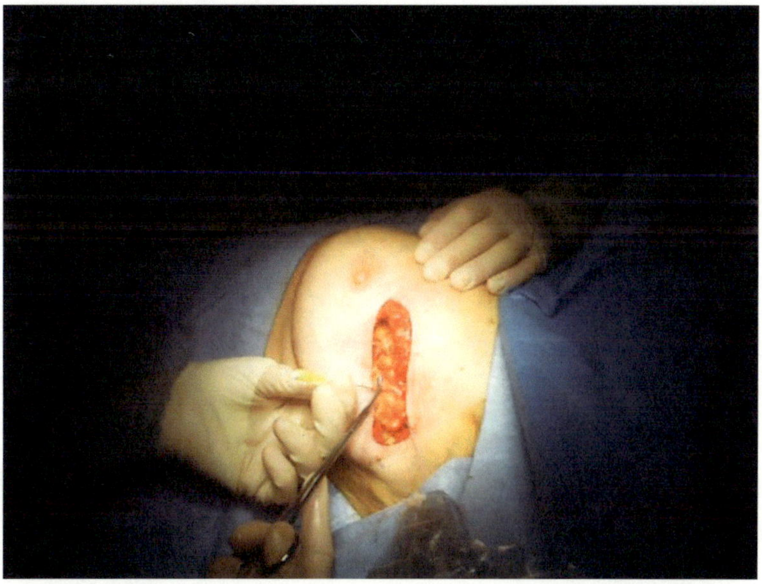

Figure 7. A graduated needle measures gland thickness

A total dose of 18 Gy or 21 Gy is delivered directly to the mammary gland depending on tumour volume according to the pilot study that is ongoing in our Institute.

In-vivo dosimetry may be useful for the optimization of the dose delivered in IORT. This optimization can help to reduce unnecessary large over-dosage regions and allows introducing reliable action levels for in-vivo dosimetry [18].

The sterile polymethyl methacrylate (Perspex; Hitesys SpA, Aprilia, Italy) collimator of the linear accelerator (LINAC) is introduced through the skin incision and placed directly over the breast target, on the tumor bed (Fig.8).

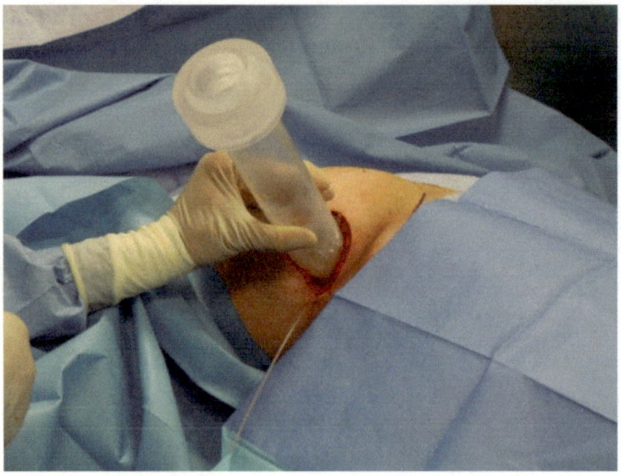

Figure 8. The collimator placed on the gland

The diameter of the collimator must be chosen according to tumor size and breast volume. As an area of 4-5 cm around the tumour has to be irradiated, the most useful collimators are 5 and 6 cm in diameter. An involuntary herniation of the gland into the collimator must be avoided, as it could result in an increased delivered dose to the superficial part of the target. The radiation technologist leads the LINAC into the surgical room; then the connection to the distal part of the applicator is performed to start the dose delivery (Fig. 9).

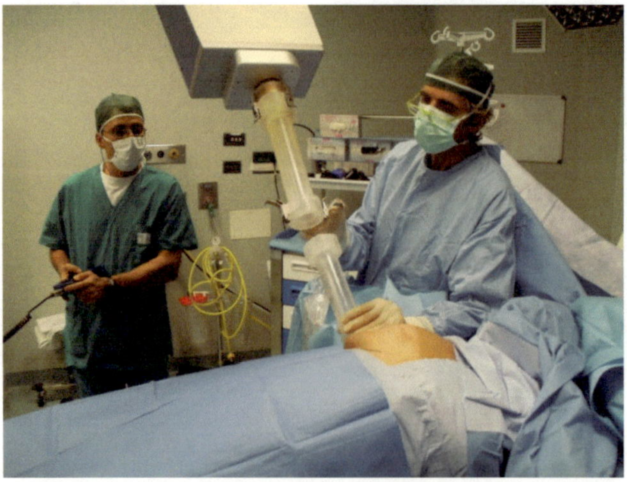

Figure 9. The radiation technologist and the surgeon lead the LINAC to connect to collimator

After collimator positioning, a series of mobile barriers are positioned around the operating table to provide a good shielding of X-rays scattered by the patient; the team then leaves operative room and the irradiation starts (Fig.10).

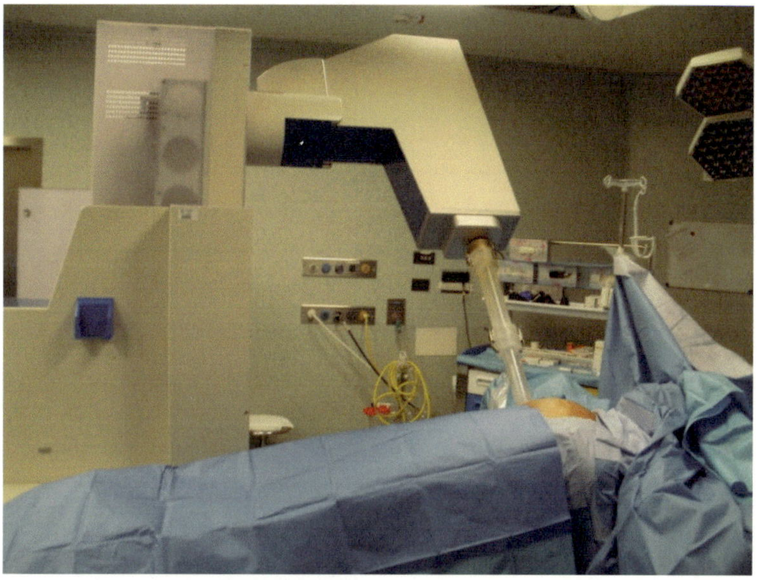

Figure 10. Operative room during radiation therapy

The prescribed dose is given in two steps; by reading the dose delivered during the first step, medical physicist and radiation oncologist can correct the second dose for giving the prescribed dose. The duration of radiation procedure is 2-4 minutes.

The applicator is immediately removed from the surgical field and the LINAC is placed far from the operating table. The shielding disc is removed and mammaplasty proceeds as the preoperative design. A contra-lateral similar reduction technique is concomitantly performed when required.

After three months a simple scale is administered to the patient and to the surgeon, separately, to evaluate post-operative aesthetic result through 4 different values: poor, sufficient, good, excellent. The surgeon's assess about aesthetic outcome results from various aspects: good proportion between size / shape of the gland and chest wall; position of areola-nipple complex on the breast meridian; distance between areola-nipple complex and infra-mammary fold [17]

Our experience as regards: age of patients (mean 61.4; median 63), TNM, amount of removed tissue, type and side of mammaplasty and aesthetic final result, is shown in Table 3.

In 9 patients mammaplasty has been performed on one side, while in the remaining 7 the surgical procedure has been bilateral (Fig. 11, 12, 13, 14, 15, 16); the choice, discussed with the patient during the preoperative consultation, has been usually motivated by psychological reasons.

Age	pTNM	Removed Tissue	Side	Mastoplasty	Patient judgment	Surgeon judgment
50	pT1c/G2-pN0-pMx	50 g	Bilateral	inf-med pedicle	excellent	excellent
61	pT1a/G2-pNx-pMx	55 g	Monolateral	inf-med pedicle	excellent	good
48	pT1c/G1-pN0-pMx	55 g	Monolateral	amputation-graft	excellent	good
49	pT1c/G2-pN0-pMx	27 g	Monolateral	inf pedicle	excellent	good
78	pT1c/G2-pN0-pMx	122 g	Monolateral	inf pedicle	good	good
67	pT1c/G3-pN0-pMx	58 g	Bilateral	inf pedicle	excellent	excellent
68	pT1b/G2-pN0-pMx	60 g	Bilateral	inf pedicle	good	good
53	pT0/G0-pN0-pMx	35 g	Monolateral	inf pedilce	excellent	good
52	pT1c/G2-pN0 pMx	41 g	Monolateral	inf pedicle	good	good
70	pT1c/G2-pN0-pMx	40 g	Monolateral	inf pedicle	excellent	good
72	pT2/G2-pN0-pMx	105 g	Monolateral	superior pedicle	excellent	good
65	pT1a/G2-pN0-pMx	23 g	Monolateral	inf pedicle	excellent	excellent
60	pT1c/G2-pN0-pMx	32 g	Bilateral	inf pedicle	excellent	excellent
49	pT1b/G1-pN0-pMx	44 g	Bilateral	inf pedicle	excellent	excellent
72	pT1b/G3-pN0-pMx	40 g	Bilateral	Inf pedicle	excellent	excellent
69	pT1c/G3-pN0-pMx	320 g	Bilateral	superior pedicle	good	good

Table 3.

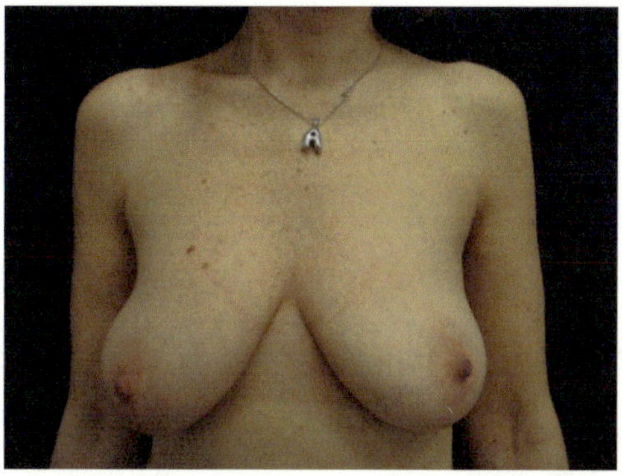

Figure 11.

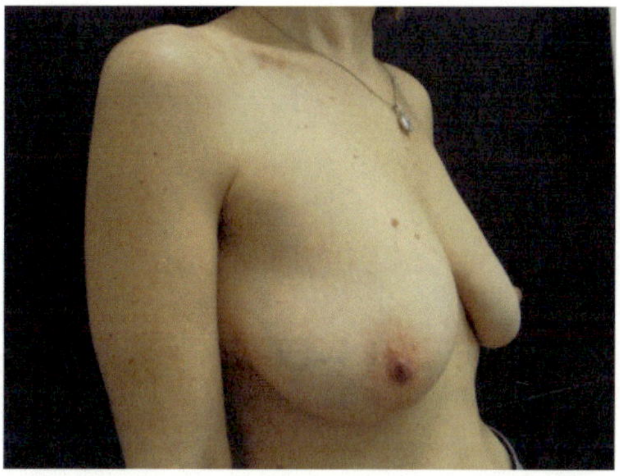

Figure 12.

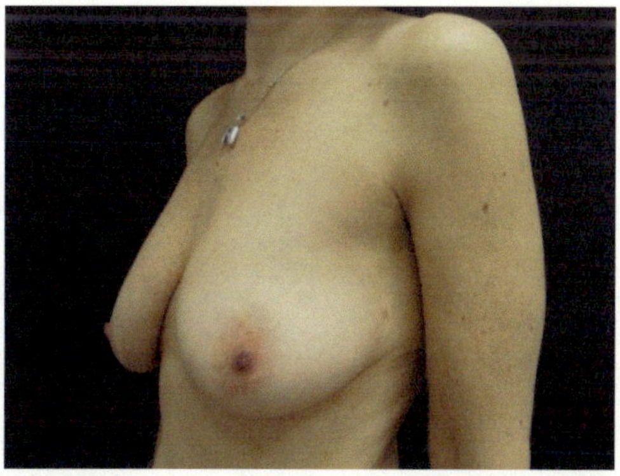

Figure 13. Preoperative view of 50-year old patient pT1c/G2-pN0-pMx ductal carcinoma on the left side; a bilateral inferior pedicle mammaplasty was performed and 50 grams of glandular tissues was removed from each breast

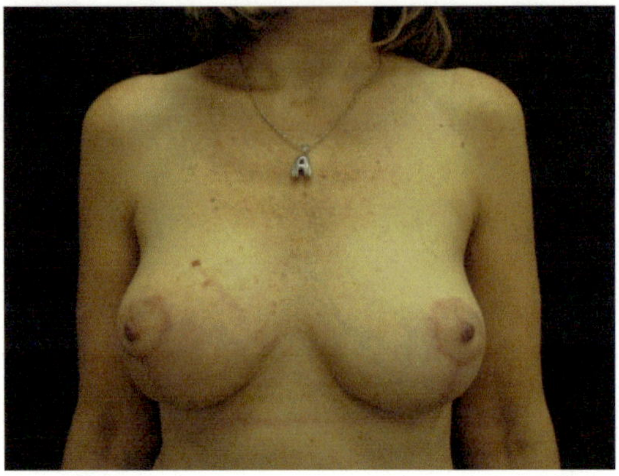

Figure 14.

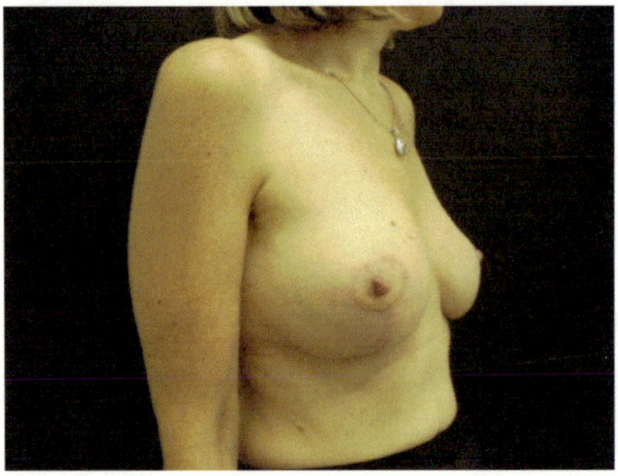

Figure 15.

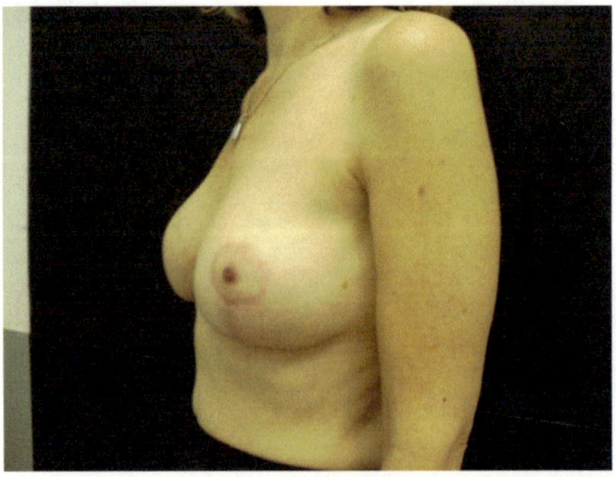

Figure 16. 6-month postoperative view of the same patient

5 of 9 patients with unilateral mammaplasty have required the contra-lateral procedure afterwards (Fig. 17, 18, 19).

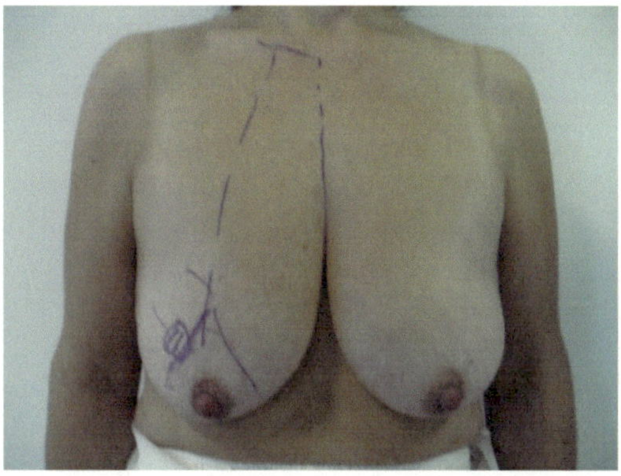

Figure 17. Preoperative view of 49-year old patient with pT1c/G2-pN0-pMx ductal carcinoma on the right side. An inferior pedicle mammaplasty was performed and 27 gram of glandular tissue was removed from right breast.

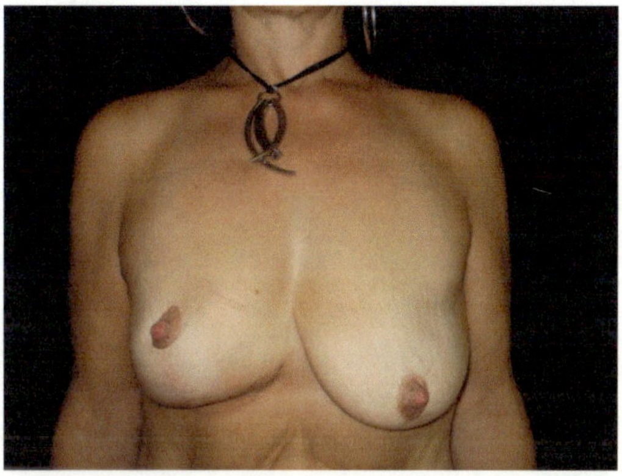

Figure 18. 6-month postoperative view of the same patient

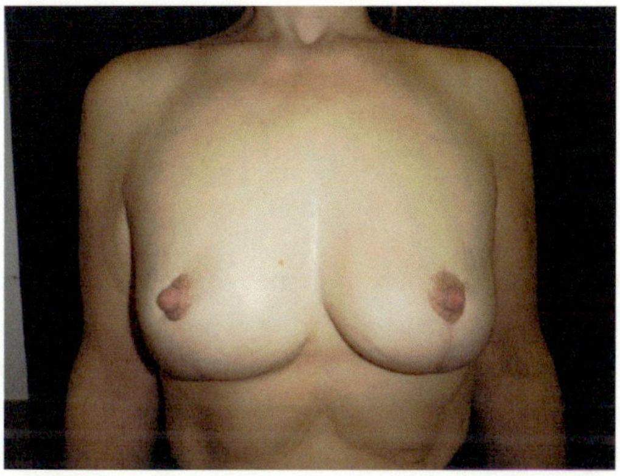

Figure 19. The same patient required reduction mammaplasty on the left side. The same procedure was applied. Final result at 3 months.

The complications observed were: areola-nipple partial loss <25% (1 case) (Fig. 20), liponecrosis (3 cases), partial vertical scar dehiscence (2 cases).

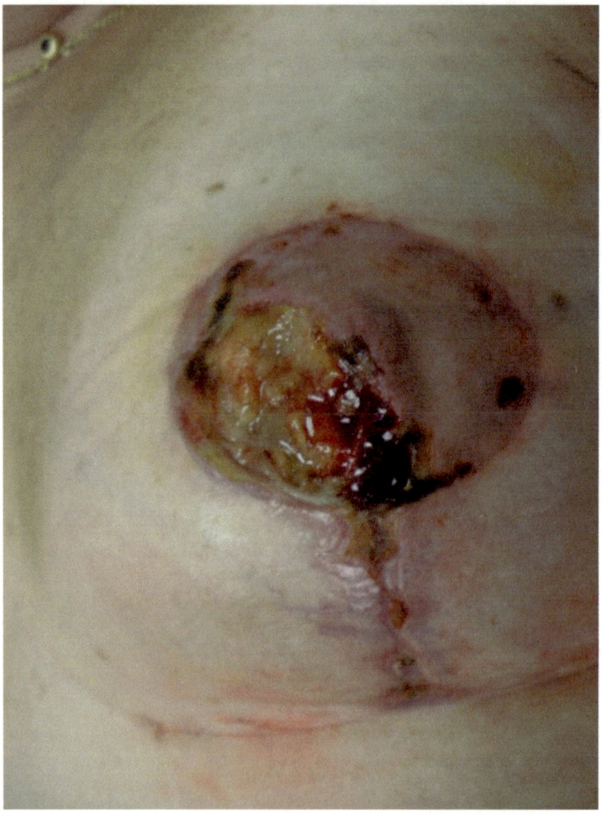

Figure 20. Skin necrosis with partial loss of areola in a 60-year old patient.

Liponecrosis caused discomfort to one patient with local oedema and erythema, but resolved spontaneously without breast deformity, while in 2 patients liponecrosis required surgical debridment in local anaesthesia. Partial areola-nipple loss is made up in local anaesthesia with skin graft from internal part of thigh and the vertical scar dehiscence improved with a revision in local anaesthesia. So far there is no evidence of local recurrence after 9 month medium follow-up (range 6-22 months). The weight of specimen is variable from 23 g to 320 g (mean g 71.13; median g 50). The postoperative result is good for 4 patients and excellent for 12, while the surgeon judged the result good in 10 and excellent in 6. No poor result has been observed.

5. Conclusion

Oncoplastic surgery represents an established new alternative in conservative breast surgery.

Several advantages of mammaplasty plus IORT have been elaborated along with the cautions to be administered during the surgery, as already discussed in our previous preliminary report [19].

The combination between glandular flaps resulting in minimal displacement and IORT, applied on cancer site immediately after lumpectomy and before mobilizing flaps, allows the optimization of irradiation effects and makes unnecessary the use of metal clips, used previously to detect the tumour bed.

The preoperative planning between oncologic surgeon and plastic surgeon is important to understand how much tissue has to be removed and which glandular pedicles can be spared, including the undermining of glandular flaps due to the positioning of a 6-8 cm diameter disk for radiation-therapy. All these aspects must be carefully discussed before projecting reduction mammaplasty in order to reduce postoperative complication such as liponecrosis and areola-nipple necrosis. If the lump is very close to nipple-areola complex, it appears safer, for oncologic and aesthetic reason, to remove it and address a contra-lateral graft.

Postoperative breast shape is evaluated by the patients and the surgeons as good/excellent. The aesthetic results have been, on an average, less satisfactory for the surgeons, especially in cases where unilateral procedure was performed.

It is maybe noteworthy to recognize that a greater number of patients over 68 found the result good, while younger patients found it excellent.

On oncologic viewpoint, we believe that the technique is reliable as long as various principles are respected: the very rigorous criteria of inclusion and the width of margin applied (> 1 cm around the tumour).

On this subject, many other Authors published data about the weight of histological specimen in oncoplastic surgery that are on average greater than those obtained after lumpectomy [20, 21].

Actually, we could verify that in another similar series of patients who were subjected to lumpectomy at the our Department the medium weight of specimen was 24.5 (median 19).

Finally, the wider extension of cutaneous access in comparison with classic lumpectomy plus IORT, permits to use easily larger disks and to apply radiation-therapy on wider extension of residual glandular flaps so as to optimize the procedure.

The technique described is reliable as long as very strict criteria of inclusion is applied. Close collaboration between surgical oncologist, plastic surgeon and radiotherapist is essential in preoperative planning and during surgery in order to obtain adequate tumour resection and good aesthetic result and to minimize postoperative complications.

The excellent cosmetic results and patients satisfaction encourage us to continue on this way, particularly in younger patients.

Author details

Simonetta Franchelli, Michela Massa and Pierluigi Santi
Chirurgia Plastica, IRCCS San Martino- IST, Istituto Nazionale per la Ricerca sul Cancro, Genova - Italy

Marina Guenzi, Renzo Corvò and Giorgia Timon
Oncologia Radioterapica, IRCCS San Martino- IST, Istituto Nazionale per la Ricerca sul Cancro, Genova - Italy

Paolo Meszaros, Davide Pertile and Ferdinando Cafiero
Chirurgia Oncologica, IRCCS San Martino- IST, Istituto Nazionale per la Ricerca sul Cancro, Genova - Italy

Acknowledgement

We thank Valentina Roberto for the drawings and Simone Callegari, MD for technical support during manuscript preparation. A special acknowledgment goes to all staff of the operative room, anesthesiologists and nurses, that supported us in all moments.

6. References

[1] Veronesi U, Cascinelli N, Mariani L et al. (2002) Twenty-year follow-up of a randomized study comparing breast-conserving surgery with radical mastectomy for early breast cancer. N Engl J Med 347(16): 1227–1232.

[2] Darby S, McGale P, Correa C et al Early Breast Cancer Trialists' Collaborative Group (EBCTCG) (2011) Effect of radiotherapy after breast-conserving surgery on 10-year recurrence and 15-year breast cancer death: meta-analysis of individual patient data for 10,801 women in 17 randomised trials. Lancet 12; 378(9804):1707-16

[3] Law TT, Kwong A. (2009) Surgical margins in breast conservation therapy: how much should we excise? South Med J. 102(12): 1234-7.

[4] Yarnold J, Bentzen S, Coles C and al (2011) Hypofractionated whole-breast radiotherapy for women with early breast cancer: myths and realities. Int J Radiat Oncol Biol Phys. 79(1): 1-9

[5] Rosen PP, Fracchia AA, Urban JA. (1975) Residual mammary carcinoma following simulated partial mastectomy. Cancer 35: 739–47

[6] Holland R, Veling SHJ, Mravunac M, Hendriks JHCL. (1985) Histologic multifocality of Tis, T1–2 breast carcinomas: implications for clinical trials of breast conserving surgery. Cancer 56:979–90

[7] Clarke DH, Lé MG, Sarrazin D, et al. (1985) Analysis of local regional relapses in patients with early breast cancers treated by excision and radiotherapy. Experience of the Institute Gustave Roussy. Int J Radiat Oncol Biol Phys 11: 137–45

[8] Vicini FA, Arthur DW. (2005) Breast brachytherapy: North American experience. Semin Radiat Oncol 15: 108–15

[9] Smith DB, Arthur DW, Buchholz TA, et al. (2009) Accelerated partial breast irradiation consensus statement from the American Society for Radiation Oncology (ASTRO). Int J Radiat Oncol Biol Phys 74: 987-1001

[10] Polgár C, Van Limbergen E, Pötter R et al (2010) Patient selection for accelerated partial-breast irradiation (APBI) after breast-conserving surgery: Recommendations of the Groupe Européen de Curiethérapie-European Society for Therapeutic Radiology and Oncology (GEC-ESTRO) breast cancer working group based on clinical evidence. Radiother Oncol. 94(3): 264-73

[11] Offersen BV, Overgaard M, Kroman N et al (2009) Accelerated partial breast irradiation as part of breast conserving therapy of early breast carcinoma: A systematic review. Radiother Oncol. 90(1): 1-13

[12] Vaidya JS, Joseph DJ, Tobias JS et al (2010)Targeted intraoperative radiotherapy versus whole breast radiotherapy for breast cancer (TARGIT-A trial): an international, prospective, randomised, non-inferiority phase 3 trial. Lancet 376(9735): 91-102

[13] Orecchia R, Leonardi MC (2011) Intraoperative radiation therapy: is it a standard now? The Breast 20 S3, 111–115

[14] Veronesi U, Orecchia R, Luini A et al (2010) Intraoperative radiotherapy during breast conserving surgery: a study on 1,822 cases treated with electrons Breast Cancer Res Treat 124: 141–151

[15] Leonardi MC, Maisonneuve P, Mastropasqua MG et al (2012) How do the ASTRO Consensus Statement Guidelines for the Application of Accelerated Partial Breast Irradiation Fit Intraoperative Radiotherapy? A Retrospective Analysis of Patients Treated at the European Institute of Oncology. Int J Radiat Oncol Biol Phys. 13

[16] Rees and La Trenta (1992) Aesthetic Plastic Surgery, W.B. Saunders Company pp. 926-987

[17] Robert Goldwyn, (1990) Reduction Mammaplasty, Little, Brown Company

[18] Agostinelli S, Gusinu M, Cavagnetto F et al (2012) On-line optimization of intraoperative electron beam radiotherapy of the breast Radiother Oncol. 17

[19] Franchelli S, Meszaros P, Guenzi M, Corvó R, Pertile D, Massa M, Belgioia L, D'Alonzo A, Cafiero F, Santi P. (2011) Preliminary experience using oncoplastic techniques of reduction mammaplasty and intraoperative radiotherapy: report of 2 cases. Aesthetic Plast Surg. 35(6): 1180-3.

[20] Masetti R, Di Leone A, Franceschini G, et al (2006) Oncoplastic techniques in the conservative surgical treatment of breast cancer:an overview. Breast J 12:174-80

[21] Clough KB, Lewis JS, Couturaud B, Fitoussi A, Nos C, Falcou MC. (2003) Oncoplastic techniques allow extensive resections for breast-conserving therapy of breast carcinomas Ann Surg. 237(1): 26-34.

Chapter 10

Effects of Radiotherapy on Pharyngeal Reconstruction After Pharyngo-Laryngectomy

Jimmy Yu-Wai Chan and Gregory Ian Siu Kee Lau

Additional information is available at the end of the chapter

http://dx.doi.org/10.5772/56604

1. Introduction

After adequate extirpation of tumours in the hypopharynx, the defect created should be reconstructed appropriately to provide optimal function. Local cervical skin flaps were first described and employed 60 years ago for the reconstruction of defects after pharyngo-laryngectomy [1]. However, the procedures involved had to be carried out in stages, which typically required 4 – 6 months for completion. In addition, the subsequent ability to swallow was frequently limited by the stenosis at the anastomotic junction. Nowadays, with more advanced techniques, reconstruction of the circumferential hypopharyngeal defect is nearly always performed at the time of resection as a single stage procedure [2]. The most commonly employed reconstructive options include the use of myocutaneous flaps (pedicled pectoralis major flap or free anterolateral thigh flap) with the skin island sutured and fashioned as a tubular conduit, or the free visceral flap (free jejunal flap).

Radiation therapy has become an integral part of treatment for malignancies in the head and neck region. Apart from being the primary treatment for radiation sensitive cancers, or as part of the organ-preserving therapy for early cancer of the larynx and hypopharynx, it is more commonly used as adjuvant treatment after surgery for advanced stage cancers. Studies showed that the 5-year survival rate as well as the loco-regional tumour control was significantly improved after combined surgical resection and post-operative radiation therapy, compared with single modality treatment.

Ionizing radiation, however, is not without side effects. In this chapter, we will focus on the effect of radiation on the functional and oncological outcome on pharyngeal reconstruction after pharyngo-laryngectomy for carcinoma of the hypopharynx.

2. Problem statement

- To investigate the effect of radiotherapy on the functional outcome of pharyngeal reconstruction
- To investigate the effect of radiotherapy on the oncological outcome of pharyngeal reconstruction
- To formulate the choice of reconstructive options for pharyngeal defects

2.1. Application area

- Reconstruction of the defect created after circumferential pharyngectomy for malignancies in the region of the hypopharynx and cervical esophagus

2.2. Methods

We identified all consecutive patients undergoing reconstruction for circumferential pharyngeal defects after resection of tumours of the hypopharynx and the cervical esophagus over a 30-years interval from 1980 to 2009. All the operations were performed in a single, tertiary referral, university based hospital. The patients' data were prospectively collected in the head and neck cancer database of the Division of Head and Neck surgery, including patients' demographic data, types of cancer treated, the operations performed, outcomes of surgery, dose of radiation given and the subsequent follow up information.

During the study period, we had performed circumferential pharyngectomy for 202 patients suffering from tumours involving the pharyngeal region. Those with tumours requiring pharyngo-laryngo-esophagectomy and subsequent gastric pull-up were excluded from the study. All patients had pre-operative work-up for tumour staging, anaesthetic assessment and nutritional build-up if necessary. All of the patients received total laryngectomy and circumferential pharyngectomy. Cervical lymph node dissection was performed if there was evidence of lymphatic metastasis. Intra-operative frozen section examination of the resection margins was performed to ensure microscopic clearance of disease.

The resultant pharyngeal defect in the form of a conduit between the oropharynx above and the esophagus below was reconstructed with either the pectoralis major (PM) flap, free anterolateral thigh (ALT) flap or free jejunal flap as described below.

3. Myocutaneous flap

3.1. Pectoralis major myocutaneous flap

Following Ariyan's [3] publication in 1979, Withers at al [4] described the technique of folding the skin island of a PM flap into a tube for the reconstruction of circumferential pharyngectomy defects. In essence, a rectangular or trapezoid skin island was designed over the pectoralis major muscle, which was transferred together with its supplying blood vessels, namely the thoracoacromial and lateral thoracic artery, to the neck for reconstruction of the neopharynx (Figure 1). The length of the pharyngeal defect should be accurately measured

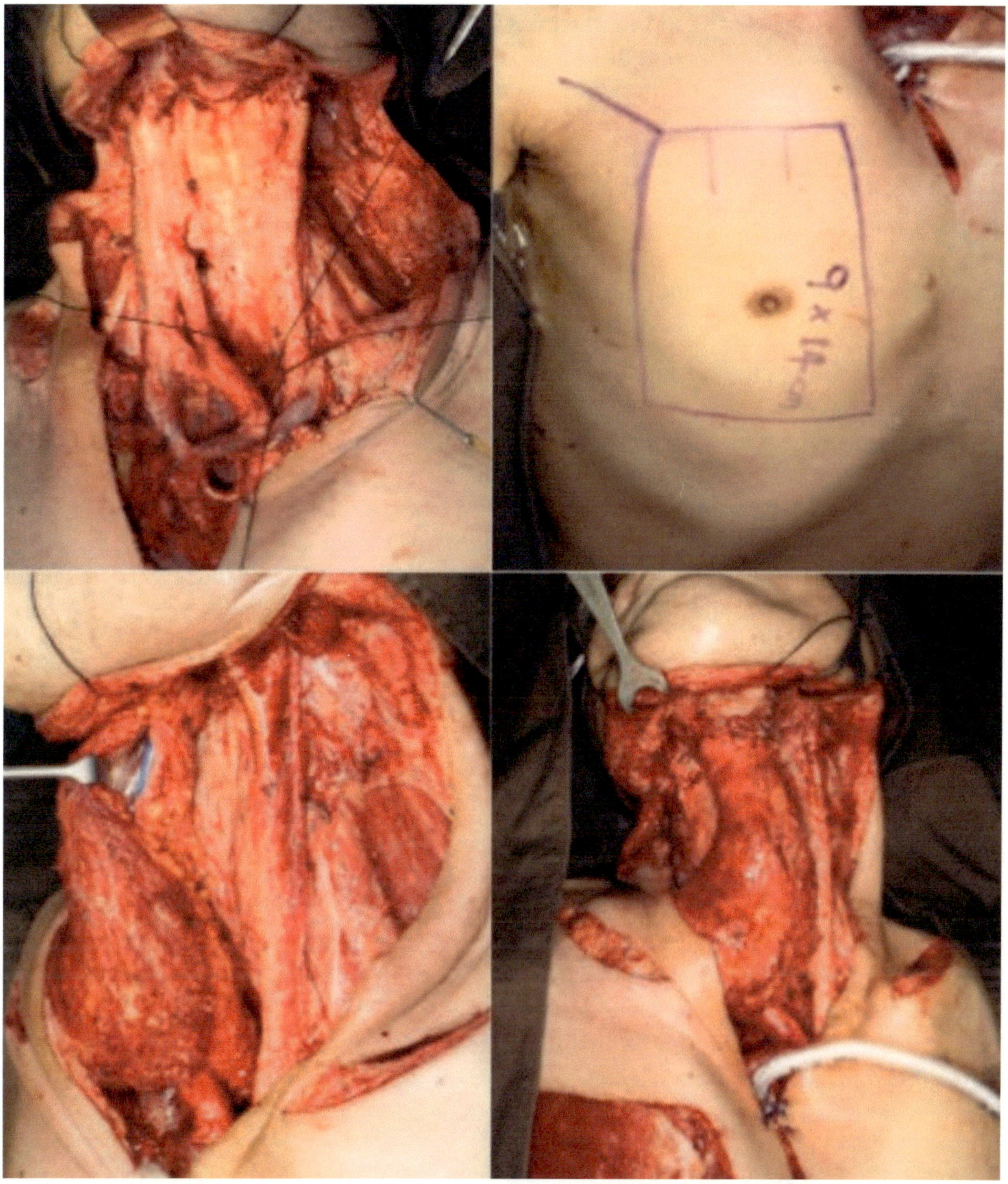

Figure 1. Circumferential pharyngeal defect reconstructed with tubed pectoralis major flap. (Above, left) circumferential defect after tumour extirpation. (Above, right) Pectoralis major myocutaneous flap with a trapezoid skin island. Slits are made at the upper edge of the skin island so that the anastomosis with the cervical esophagus can be performed in an interdigitating fashion. (Below, left) The flap is transferred to the neck through a subcutaneous tunnel. The skin island is folded into a tube and sutured to form a conduit. Nasogastric tube is inserted for feeding during early post-operative period (Below, right) Completion of inset, showing the pectoralis muscle facing outside. It was then resurfaced with autologous split thickness skin graft.

and the length of the skin island marked out accordingly. In addition to the expected 10% shrinkage in size after harvesting, the length of the skin island should be at least 2cm longer than the defect if the interdigitating suturing method is employed at the lower (esophageal) anastomosis. The width of the skin island should be at least 6cm in order to provide a reconstructed skin tube with a 2cm diameter. In order to include as many perforators as possible, the skin island should be placed entirely on the pectoralis major muscle, and the shape of the skin island should be in the form of a trapezium, with the shorter edge on its upper part. This upper edge of the skin island, when transposed to the neck, will be anastomosed to the cervical esophagus which is always smaller in caliber than that of the oropharynx.

After passing through the subcutaneous tunnel, the skin island of the pectoralis major flap is fashioned into a tube. It is then sutured to the oropharynx above and the cervical esophagus below. In order to prevent stricture formation of the circular mucocutaneous anastomosis with the esophagus, the anastomosis is carried out in an interdigitating fashion [5]. Three incisions are made at the stump of the cervical esophagus. Similar incisions are made over the skin island so that at anastomosis, skin island flaps can be sutured in an interdigitating fashion with the corresponding flaps from the cervical esophagus. The resultant anastomosis is in a wave-like form rather than circular. Upon healing, the scar stretches the anastomosis wider rather than causing it to stenose.

3.2. Free anterolateral thigh flaps

Free anterolateral thigh fasciocutaneous flap was popularized by Song *et al* in 1984 [6] and later modified by Koshima *et al* [7-8] for clinical application. As in the pectoralis major myocutaneous flap, the skin paddle of the anterolateral thigh flap can be fashioned into a tube for the reconstruction of the circumferential pharyngeal defects after tumour resection. It is supplied by the musculocuatneous or the septocutaneous perforators of the descending branch of the lateral circumflex artery and its venae commitantes. After transfer from the donor site, the blood supply of the flap is reconstituted by microvascular anastomosis of the vascular pedicle with appropriate vessels in the neck. The vascular anastomosis is performed under magnification with interrupted sutures. Any suitable branch from the external carotid artery can be used as the recipient artery, and either the external jugular vein or any branch of the internal jugular vein can be selected as the recipient vein.

4. Free visceral flap

4.1. Free jejunal flap

The free jejunal flap was first described by Seidenberg [9] in 1958. It represents one of the most popular methods for the reconstruction of circumferential pharyngeal defects nowadays (Figure 2). Through an upper midline laparotomy, the required length of the jejunum is harvested. Usually, the segment of the jejunum supplied by the second vascular arcade from the ligament of Treiz is used, as the vessels are relatively straight with a size comparable with that of the recipient vessels in the neck. The size of the jejunal lumen is comparable with that of the cervical esophagus, and an end-to-end anastomosis can usually be accomplished easily. The discrepancy in size usually occurs at the anastomosis between the oropharynx and the jejunum. It can be handled by slitting the jejunal wall at the anti-mesenteric border so that a longer circumference is made available for anastomosis.

In order to minimize the ischaemic time of the jejunal flap, the recipient vessels in the neck should be prepared and ready for microvascular anastomosis before the flap is detached from its blood supply in the abdomen. The jejuno-esophageal anastomosis is performed for fixation of the bowel, followed by vascular anastomosis. Once the perfusion of the bowel is restored, upper anastomosis between the oropharynx and the jejunal flap can be performed leisurely without time constraint.

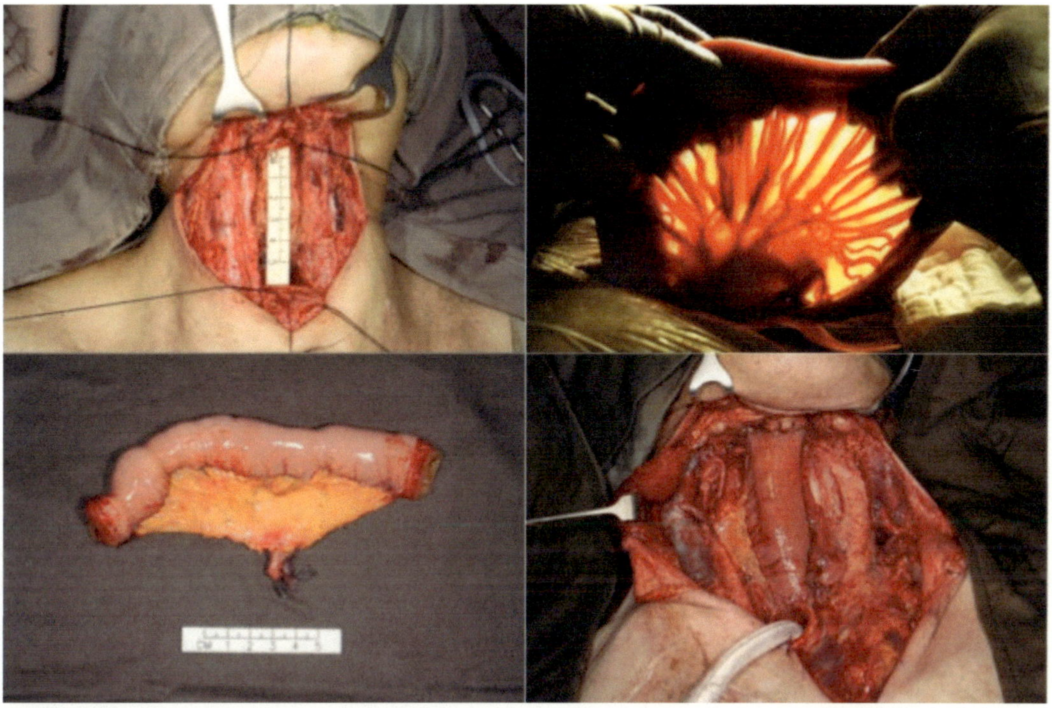

Figure 2. Circumferential pharyngeal defect reconstructed with the free jejunal flap. (Above, left) Circumferential pharyngeal defect created after resection of tumour at the hypopharynx. The length of the defect between the oropharynx above and the cervical esophagus below is accurately measured. (Above, right) Through a upper midline laparotomy, the segment of jejunum of appropriate length is harvested. The vascular anatomy is better appreciated with light shining from behind. The second arcade of blood vessels is commonly harvested because of the favorable configuration. (Below, left) The flap is delivered after division of the vascular pedicle. Effort should be made to shorten the ischaemic time as much as possible. The jejuno-esophageal anastomosis is performed first, followed by microvascular anastomosis under magnification. (Below, right) Upon completion of the upper oropharyngo-jejunal anastomosis. The anti-mesenteric border of the jejunal flap may be splitted to accommodate the larger size of the oropharynx.

Post-operative gastrograffin swallow study was performed 10 days after surgery to complete wound healing before oral feeding was resumed. For those patients who developed pharyngo-cutaneous fistula with significant leakage, a control pharyngostome was performed to allow the patient to recover from the inflammation before second stage repair was contemplated. Those minor leakages with no clinical evidence of inflammation were treated conservatively and monitored closely. Contrast swallow study was performed subsequently to confirm complete wound healing before allowing oral feeding. Those patients with significant delay of post-operative adjuvant chemoradiotherapy secondary to anastomotic leakage or necrotic flap requiring multiple surgeries were excluded. All patients were referred to clinical oncologists for post-operative adjuvant chemoradiotherapy.

External beam radiotherapy, when indicated, was commenced as soon as possible after all the surgical wounds were healed. Radiation was delivered by Cobalt 60, 4MV, or 6MV linear accelerator, which was given once per day with daily fraction size of 2Gy, 5 days per week. The spinal cord was shielded after 4500 cGy and the flap for the pharyngeal reconstruction was included in the high dose field of radiation.

All the patients were followed up regularly with clinical and endoscopic examination to detect tumour recurrence and complications such as anastomotic stricture and donor site morbidities. A stage-to-stage comparison of the loco-regional tumour control was performed between different methods of reconstruction.

5. Results

5.1. Functional results after reconstruction of circumferential pharyngeal defects

Pharyngo-cutaneous fistula leading to anastomotic leakage is a serious complication during the early post-operative period. If not detected and treated early, it will result in severe infection and carotid artery blowout. While surgical technique is important to prevent such complication, the choice of reconstructive options may also affect the incidence of early anastomotic leakage. According to our experience over the past two decades [10], the incidence of early post-operative pharyngo-cutaneous fistula was 23.9% when pectoralis major flap is used, 12.5% when the free anterolateral thigh flap is used, and 4.6% when the jejunal flap is used. Similar results are demonstrated in other series [11-12]. Majority (65.5%) of the patients with salivary leakage required exteriorization by creating a control pharyngostome pharyngostomy, followed by a second stage reconstruction after the inflammation subsided. The rest of the patients (34.5%) were managed successfully by conservative approach. There was no significant relationship between pre- or post-operative radiotherapy (p=0.848) and early post-operative fistula formation. However, among the patients who leaked, those with a history of irradiation were significantly more likely to required exteriorization to control the infection (88.2% required exteriorization vs. 11.8% managed conservatively).

One of the most important functional problems after circumferential pharyngectomy is swallowing. In addition to the restoration of the resected pharyngeal conduit for the passage of food, proper swallowing in these patients requires patent anastomosis without stenosis.

Late anastomotic stricture rate was 27.2% in the pectoralis major flap group, 12.5% in the anterolateral thigh flap group, and 2.3% in the jejunal flap group. The mean time to develop dysphagia secondary to the stricture was 18.4 months. The only factor that was found to have significant association with stricture formation was the history of pharyngo-cutaneous fistula during early post-operative period (p=0.023). The history of radiotherapy has not significantly increased the risk of anastomotic stricture.

In the group of patients who had no demonstrable anastomotic stricture, only some of them were able to resume the usual diet before operation, the proportion of which being 35.8% in the pectoralis major flap group, 38.2% in the anterolateral thigh group and 61.9% in the jejunal flap group. The rest of the patients tolerated fluid or soft diet only. The presence of mucus secretion as well as peristalsis in the jejunal flap may aid the passage of food bolus. Table 1 summarizes the functional outcome after different types of reconstruction techniques.

	Total number	Early fistula (%)	Late stricture (%)	Resume usual diet (%)	Donor site morbidity (%)
Pectoralis major flap	92	22 (23.9)	25 (27.2)	24 (35.8)	7 (7.6)
Anterolateral thigh flap	24	3 (12.5)	3 (12.5)	8 (38.1)	1 (4.2)
Jejunal flap	86	4 (4.6)	2 (2.3)	52 (61.9)	2 (2.3)

Table 1. Comparison of the functional outcome after different types of reconstruction of the circumferential pharyngeal defect.

Majority of the patients require a combined surgical resection followed by post-operative adjuvant radiotherapy for treatment of carcinoma hypopharynx. While free jejunal transfer appears to result in the lowest rate of early anastomotic leakage as well as late anastomotic stricture, the radiation tolerance of the visceral flap is always a concern. The jejunum is a radiosensitive organ in its native site in the abdomen. Radiation injury to the gastrointestinal tract is well recognized in patients with pelvic or colorectal malignancies who received external beam irradiation up to 60Gy [13]. Acute functional changes include diarrhea, bloating and abdominal pain, and reported late sequelae include chronic diarrhea, malabsorption with steatorrhoea, abdominal spasms, intestinal obstruction, bleeding and fistula formation [14]. It is observed that, apart from the total dose of radiation delivered, the volume of the small intestine exposed to the radiation is also important in determining the severity of the presenting symptoms of the patients [15]. When biopsies are taken from the mucosa of the jejunal flap before and after radiotherapy and compared under microscopy and scanning electron microscopy [16], it was found that there is generalized edema and loss of villi during early post-radiation period (Figure 3). The thinning of jejunal mucosa, the focal loss of glands and the blunting of the villi are found at 6 months after completion of irradiation and persisted throughout the following 2 years. However, none of the jejunal biopsies displayed any area of necrosis or ulceration secondary to radiation.

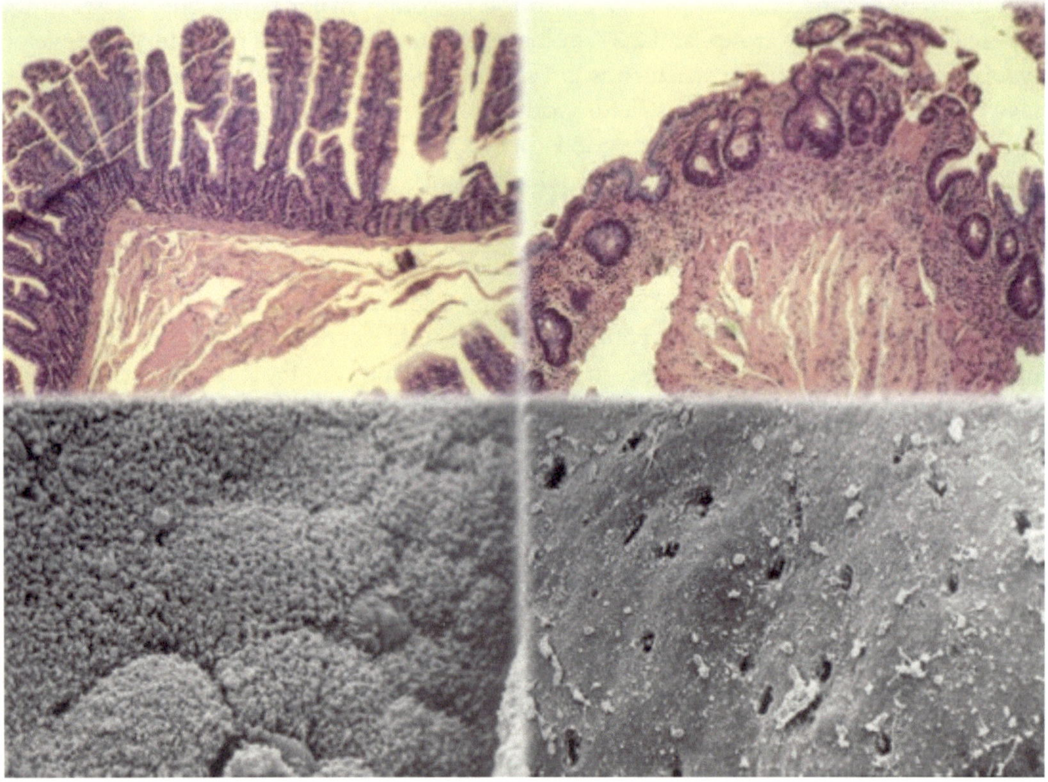

Figure 3. Effect of radiotherapy on the jejunal flap (Above, left) Haematoxylin and eosin stain, x 80. Before radiotherapy. The villi of the jejunal flap in normal configuration. (Above, right) Haematoxylin and eosin stain, x 80. 2 months after radiotherapy. Shortening and blunting of villi with increased fibrosis and retraction of the glands noted. (Below, left) Scanning electron micrograph. Before radiotherapy. Normal jejunal mucosa with healthy microvilli and cell boundaries well demonstrated. (Below, right) Scanning electron micrograph. 2 months after radiotherapy. There is generalized shortening of the microvilli. Pitted areas represent the loss of microvilli.

5.2. Oncological results after reconstruction of circumferential pharyngeal defects

Despite the increasing popularity of organ-preserving chemoradiation for early stage cancer of the hypopharynx, surgery remains the preferred therapeutic option for locally advanced disease, when post-operative radiotherapy is generally indicated. Evidence shows that adjuvant radiotherapy significantly improved locoregional control as well as survival after surgery for carcinoma at the hypopharyngeal region [17]. However, because of the potential risk of radiation damage to the jejunal flap which has been transferred to the neck as a free flap after tumour extirpation, it has been a common practice among the clinical oncologists to reduce the radiation dosage for patients who had received free jejunal flap reconstruction. Our experience showed that the mean radiation dosage given to patients after visceral flap transfer was much lower than those after cutaneous flap reconstruction (54.8Gy vs. 62.2Gy) [18].

At this level of radiation, there was no evidence of secondary ischaemia or necrosis of the flap upon completion of the adjuvant treatment. Despite the reduction of radiation dosage in the jejunal group, the difference in the 5-year actuarial loco-regional tumour control between the 2 groups of patients was not significant for TNM stage II (61% vs. 69%, p = 0.9) and III (36% vs. 46%, p = 0.2) disease. However, for stage IV disease, patients with jejunal flap reconstruction had significantly poorer loco-regional tumour control (32% vs. 14%, p = 0.04) (Table 2). This may be explained by the failure of the reduced dose of radiation to deal with the more widespread microscopic tumour deposits in the advanced stage of disease. Despite the apparently better functional outcome using the jejunal flap to reconstruct circumferential pharyngeal defects, the choice of reconstructive options should never compromise the oncological control. The surgeons should communicate with the oncologists, and if a reduced dose of radiation is to be given, then jejunal transfer should not be the first choice of reconstruction in advanced staged carcinoma of the hypopharyngeal region.

	Cutaneous flap reconstruction	Free jejunal flap reconstruction	p-value
Stage II	61%	69%	0.9
Stage III	36%	46%	0.2
Stage IV	32%	14%	0.04

Table 2. Comparison of 5-year actuarial loco-regional tumour control between patients using cutaneous flap and free jejunal flap for pharyngeal reconstruction.

6. Conclusion

Reconstruction of circumferential pharyngeal defects after tumour extirpation remains a therapeutic challenge. In order to achieve the best functional and oncological outcome, the choice of the reconstructive options should be individualized according to the patients' characteristics.

Author details

Jimmy Yu-Wai Chan* and Gregory Ian Siu Kee Lau
Division of Head and Neck Surgery, Department of Surgery
University of Hong Kong Li Ka Shing Faculty of Medicine, Queen Mary Hospital, Hong Kong SAR, China

* Corresponding Author

7. References

[1] Wookey H. The surgical treatment of carcinoma of the hypopharynx and the oesophagus. Br J Surg 1948; 35: 249 – 66.

[2] Surkin MI, Lawson W, Biller HF. Analysis of the methods of pharyngoesophageal reconstruction. Head Neck Surg 1984; 6: 953 – 70.

[3] Ariyan S. The pectoralis major myocutaneous flap. A versatile flap for reconstruction in the head and neck. Plast Reconstr Surg 1979; 63: 73-81.

[4] Withers EH, Franklin JD, Madden Jr JJ, et al. Pectoralis major musculocuatneous flap: a new flap in head and neck reconstruction. Am J Surg 1979; 138: 537-43.

[5] Lam KH, Wei WI, Lau WF. Avoiding stenosis in the tubed greater pectoral flap in pharyngeal repair. Arch Otolaryngol Head Neck Surg 1987; 113(4):428-31.

[6] Song YG, Chen GZ, Song YL, et al. The free thigh flap: a new free flap concept based on septocutaneous artery. Br J Plast Surg 1984; 37:149-59.

[7] Koshima I, Yamamoto H, Nagayama H, et al. Free combined composite flaps using the lateral circumflex femoral system for repair of massive defects of the head and neck regions: an introduction to the chimeric flap principle. Plast Reconstr Surg 1993; 92:411-20.

[8] Koshima I, Fukuda H, Ohta S, et al. Free anterolateral thigh flaps for reconstruction of head and neck defects. Plast Reconstr Surg 1993; 92:421-8.

[9] Seidenberg B, Rosenak SS, Hurwitt ES, et al. Immediate reconstruction of the cervical esophagus by a revascularized isolated jejunal segment. Ann Surg 1959; 149:162-71.

[10] Chan YW, Ng RW, Liu LH, Chung HP, Wei WI. Reconstruction of circumferential pharyngeal defects after tumour resection: reference or preference. J Plast Reconstr Aesthet Surg 2011; 64(8): 1022 - 8.

[11] Ho MW, Houghton L, Gillmartin E, et al. Outcomes following pharyngolaryngectomy reconstruction with the anterolateral thigh (ALT) free flap. Br J Oral Maxillofac Surg 2012; 50(1): 19 - 24.

[12] Yu P, Hanasono MM, Skoracki RJ, et al. Pharyngoesophageal reconstruction with the anterolateral thigh flap after total laryngopharyngectomy. Cancer. 2010; 116(7): 1718 - 24.

[13] Andreyev HJ. Gastrointestinal problems after pelvic radiotherapy: the past, the present and the future. Clin Oncol 2007; 19: 790-9.

[14] Coia LR, Myerson RJ, Tepper JE. Late effects of radiation therapy on the gastrointestinal tract. Int J Radiat Oncol Biol Phys 1995; 31: 1213-36.

[15] Fiorino C, Alongi F, Perna L, et al. Dose-volume relationship for acute bowel toxicity for patients treated with pelvic nodal irradiation for prostate cancer. Int J Radiat Oncol Biol Phys 2009; 75: 29-35.

[16] Wei WI, Lam LK, Yuen PW, Kwong D, Chan KW. Mucosal changes of the free jejunal graft in response to radiotherapy. Am J Surg 1998; 175: 44-6.

[17] Vandenbrouk C, Eschwege F, De La Rochefordière A. Squamous cell carcinoma of the pyriform sinus: retrospective study of 351 cases treated at the Institut Gustave-Roussy. Head Neck Surg 1987; 10: 4-13.

[18] Chan JY, Chow VL, Chan RC, Lau GI. Oncological outcome after free jejunal flap reconstruction for carcinoma of the hypopharynx. Eur Arch Otorhinolaryngol 2011 Nov 18 [Epub ahead of print].

Section 4

Quality of Life Issues

Chapter 11

Breast Cancer Post Treatment Quality of Life

Sabina Saric

Additional information is available at the end of the chapter

http://dx.doi.org/10.5772/56041

1. Introduction

Malignant tumors are one of the biggest medical as well as social problems worldwide. The number of patients suffering from malignant diseases has doubled in the last 30 years. Therefore, malignant diseases are a global problem among which breast cancer is the most common malignant tumor in women (25% out of all malignant tumors). The diagnoses of cancer as well as functional disability due to therapeutic interventions affect life quality of breast cancer patients. Effects of therapeutic interventions, viz.lymphoedema, limited shoulder movement, along with socio-psychological effects, play a major influence on these patients, but these topics have not received any significant attention.

Due to lack of National Cancer Registry as well as precise data, within Bosnia and Herzegovina, it is still impossible to determine the exact number of breast cancer patients. The estimated number of breast cancer patients can only be assessed based on the World malignant illnesses database along with average malignant illnesses occurrence within the European region. Based on literature, 1373 new breast cancer patients can be estimated on annual basis in Bosnia and Herzegovina. (GLOBOCAN 2000. Cancer Incidence, Mortality and Prevalence Worldwide. Version 1.0)

The total number of registered malignant illnesses patients within the Federation of Bosnia and Herzegovina in 2007 was 4. 147. Within the Federation of Bosnia and Herzegovina a total of 487 new breast cancer patients were registered, with 98.2% (478) females and 1.8% (9) males. [1] n 148/487 were in the age group 45-54 years followed by 132/487 in 55-64 years and 125/487 breast cancer patients belonged to age group above 65 years of age. [1]

Based on Canton Sarajevo malignant illnesses database, incidence of breast cancer ranks accounts for a total of 181cases (177 women, 4 men) [1]

In the same year, in Republic Srpska from 4.268 registered malignant cancer patients, 8.83% were patients of breast cancer, (409) with 397 women and 12 men.

In Brcko District no malignant diseases registry exists. The number of affected patients is roughly estimated from the District hospital reports. Information as such is deemed unreliable and therefore invalid.

Due to lack of cooperation between the two entities and Brcko District the total number of malignant diseases in Bosnia and Herzegovina does not exist.

Quality of life is defined as individual perception of his or her own position of existence based on two contexts. Namely, within cultural context as well as within personal assessment system context of one`s existence in relation to one`s expectations, goals, standards and priorities. This concept is truly broad as its complexity encompasses individual physical health, psychological status along with social and economic circumstances surrounding each individual. Quality of life reflects individual perception in relation to the level by which their needs have been met as well as the level by which their happiness has been achieved, personal satisfaction etc., regardless of their physical health status and their social and economic circumstances. Quality of life at its simplest definition would pose the following question, `How good is my life for me`. Life quality enhancement truly necessitates nourishment of human power and capability that include; optimism, creativity, inner motivation and sense of responsibility for oneself as well as for close ones. This is a basic human principle which projects the idea that the life we have is the life worth living.

Breast cancer patients therefore must have a realistic view on life and be aware of all possibilities. Their life quality can be affected not only by physical and functional deficit caused by the malignant illness, but also by medical intervention effects which include therapeutical interventions that significantly impair their physical as well as psychological life quality during and after the treatment. The major effects of medical therapeutical interventions are limited shoulder movement of the operated arm-side, lymphedema on the operated arm-side as well, and socio-psychological effects which impair emotional ability to function within social context.

As the treatment takes course, effects mentioned above take root. In line with that interdisciplinary cooperation for efficient treatment and rehabilitation where the breast cancer patient is the focus becomes of immense importance. It is also important to mention that animation of patients' interest and will to fight is of utmost significance. However, the lack of maintenance and the post treatment life quality downfall among breast cancer patients has failed to draw any significant attention of health professionals in Bosnia and Herzegovina. Causes of this situation are many, namely health professionals are overburdened by new cases to the extent that follow-up treatments and consultations happen only in cases of illness residue. Needless to say this situation leaves absolutely no space for adequate breast cancer post treatment, life quality consultation and rehabilitation.

Early detection and diagnosis of medical intervention effects during breast cancer treatment, offers higher treatment success and better control possibilities. [3] Patient education on preventive measures of post-treatment effects as well as medical treatment effects during the course of treatment itself should be included in the breast cancer treatment protocol in order to better enhance the quality of breast cancer patients lives.

Taking into consideration that in Bosnia and Herzegovina team work between health professionals and breast cancer patients during the course of an early stage protocol for prevention and rehabilitation, is not included as a standard protocol procedure, it is important to note that proposal of such inclusion is a crucial factor in establishing a continuum for quality of life to prevent possible side effects. Breast cancer patients' quality of life estimates based on standardized questionnaire of EORTC (The European Organization for Research and Treatment of Cancer QLQ-C30 and Module BR23) [4] enables an efficient overview of the proposed treatment protocol.

2. Goals

Goals of this research are breast cancer patients' quality of life though post treatment period, all estimates are based on standardized questionnaire of European Organization for Research and Treatment of Cancer (hence onwards referred to as EORTC) QLQ-C30 and Module QLQ-BR23 (version 3). This study will help develop a protocol, to prevent breast cancer patients medical treatments side effects by offering an early intervention.

3. Material and methods

100 patients were selected from a Sarajevo Canton Breast Cancer Patients Association (Renesansa) previously treated for breast cancer. Research sample selection was based on three criteria; a)patients had completed their treatment between 1-20 years prior to enrollment in the study b) patients had not experienced a relapse c) patients had not experienced infection. All patients were examined and completed the quality of life questionnaires QLQ-C30 and Module QLQ-BR23. These questionnaires are standardized by the European Organization for Research and Treatment of Cancer (EORTC) – The Quality of Life Unit and consist of 30 questions which include the following: [5, 6]

1. Five functional scales: physical, role, emotional, social, and cognitive.
2. Three scales of symptoms: fatigue, pain, nausea, and vomiting.
3. Scale of global health status / quality of life

QLQ-C30 includes a string of single items which estimate additional symptoms that are often the subject of complaints of breast cancer patients: dyspnea, loss of appetite, insomnia, constipation, diarrhea and experiencing financial burden of the disease.[5, 6]

QLQ-BR 23 consists of 23 questions and includes the following:

1. Five multiple point scales: self-body image, sexual functioning, sexual satisfaction, future perspectives, and scales of symptoms that encompass systematic therapy side effects, breast symptoms, arm symptoms, disturbance due to hair loss.[5, 6]

Questionnaire result interpretation was carried out according to EORTC scoring manual.[5, 6] The patients were divided into a) limited shoulder movement group (39%), b) lymphoedema group (49%), c) both (27%), d) patients who experienced no effects (39%).

Correlation between risk-factors (BMI (body mass index), smoking, alcohol consumption and physical activities) and lymphoedema occurrence was calculated. In addition a correlation between above mentioned risk-factors and scale of physical performance was also calculated. The quality of life differences amongst the patients, who underwent radical mastectomy and partial mastectomy, were also examined. Moreover the correlation among breast cancer patients who manifested side effects of a given medical treatment in contradistinction to those who did not experience any of the side effects was assessed.

4. Results

Average age of study group was 58 (±1) years. 76% (n=100) of patients fall into the age group of over 50 years of age, while 38% belonged to the age group 51 – 60 years and 3% were <40 years of age. The study group included 54% retired, 31% employed and 15% unemployed patients. Work absence analysis shows 64% employed patients (n=31) had to take leave from 7 to 12 months from work for illness related reasons while 23% had leave of absence more than a year. A total of 48% patients (n=54) retired due to breast cancer.

According to the type of surgical treatment 71% patients underwent radical mastectomy with axillary dissection and 29% underwent partial mastectomy with axillary dissection. The nonexistence of National Program of Early Breast Cancer Detection adds to this number. In countries that have and practice early breast cancer detection programs around 60% of breast cancers are diagnosed at the early stage and have a 5 year survival up to 95%. [7, 8]

The study results show that there was a statistical significance in the difference in the quality of life between patients undergoing radical mastectomy as compared to partial mastectomy with partial mastectomy patients reporting a better quality of life. Thus indicating that, whenever possible, partial mastectomy should be prioritized in order to save and preserve QOL of breast cancer patients with reference to their body self-image.

None of the patients undergoing radical mastectomy received a breast reconstruction which could restore the body self-image, a key indicator for a better quality of life among the breast cancer patients. Total rehabilitation of breast cancer patients should include an option of reconstructive surgery and should be offered to breast cancer patients as a protocol.

72% received chemotherapy post operatively in the protocol, 79% had local radiotherapy (breast or chest wall), 76% received axillary and supraclavicular radiotherapy and 57% had hormonal therapy.

There were 61% overall survivors at 5-year 23% for 6-10 years, 9% at 11 to 15 years and 7 % for 16-20 year period. As compared to Bray F. et. al. [8] this study shows a lesser survival after 6 years post treatment period.

Body mass index (BMI) indicated that 72% patients were obese while only 28% had normal body weight. Variation range of BMI was between 22-49 kg/m^2. Average BMI value was 28.1kg/m^2 which indicated obesity, while in the research group of arm lymphoedema (n=49) average BMI value was 29.9 kg/m^2 which pointed to higher numerical value obesity.

Considering that World Health Organization lists obesity as postmenopausal breast cancer risk factor, it is recommended that breast cancer protocols must include compulsory body weight control along with nutritional programs and advice by a nutritionist. Several studies researched obesity as a lymphoedema risk factor. Memorial Sloan Kettering Cancer Center [9] study showed that BMI is a significant predictive lymphoedema risk factor among patients who received radiation and surgical treatment.

From the total number of research patients (n=100) preordained physical activity of minimum 30 minutes per day was done by 23% of research patients, while 77% followed sedentary lifestyle. Within its report on health conditions in 2002, the World Health Organization mentions physical inactivity as a breast cancer risk factor. [10] Sedentary way of life causes loss of energy that directly causes lack of care for one's physical and social condition thus negatively affecting breast patients' quality of life.

From the total number of research patients 34% were smokers and 77% suffered from fatigue while 60% suffered from breathing difficulties in from of dyspnea, thereby leading to a poor quality of life.

6-12% patients consumed alcohol either regularly or sporadically. Regular alcohol consumption increases breast cancer risk: two drinks a day by 30-40% and this is most probably due to increased level of estradiol. [11]

Risk factor correlation: BMI, smoking habit, alcohol consumption and physical activities with lymphoedema occurrence indicated the existence of statistical significance ($p<0,05$) only between BMI and physical activity. This enhances the need for body weight control along with physical activities program with the aim to prevent lymphoedema-medical-intervention side effects. Risk factors correlation: BMI and physical activities and physical scale, point out statistical significance on the level 0, 05 with 2 plasticity degrees, while between smoking habits and alcohol consumption no statistical relevance was noted.

According to medical intervention side effects, 39% experienced limited range of shoulder movement on the operated side, 49% experienced arm lymphoedema (the operated side arm circumference is bigger by 2cm or more than the circumference of the unaffected arm), 27% experienced lymphoedema and limited range of shoulder movement among which is one patient with affected plexus brachial on the operated side along with one patient with breast lymphoedema while 39% of patients experienced no side effects.

This study has shown the explicit statistical significance ($p<0, 05$) between breast cancer patients quality of life differences with medical intervention side effects (lymphoedema and limited range of shoulder movement) in relation to quality of life of breast cancer patients that experienced no medical intervention side effects. As was to be expected, this research has shown that medical intervention side effects negatively affect life quality of researched patients, hence it can be concluded that medical intervention side effects prevention program is of immense importance and must be included in the breast cancer treatment protocol itself.

Within sample group, lymphoedema manifestation in relation to the type surgery, from the total number of research patients with lymphoedema (n=49) 68% of patients underwent radical mastectomy and axilla dissection, while 32% of patients underwent partial mastectomy and axilla dissection.

In relation to radiotherapy treatment, 84 %(n=49) patients had received local and lymph area radiotherapy and developed lymphedema, while 16% belonged to the group which did not receive radiotherapy treatment. The more radical the surgery and area of radiotherapy, the bigger the chance of medical intervention side effects occurrence.

Average age within the lymphoedema research group of patients was 59, 8 years of age while the average number within the group of patients that experienced medical intervention side effects were 56, 5 years of age. Kiel and Rademacker [12] reported that age of breast cancer patients is a statistically speaking significant factor which leads to lymphoedema.

All data point out to the significance and the need of early rehabilitation treatment related to the basic illness. In this research, on the other hand, only 5% of the patients underwent early rehabilitation treatment, 13% underwent delayed rehabilitation treatment related to medical intervention side effects, while 82% did not undergo physical treatment related to the breast cancer. This is to be noted as lack of treatment protocol for breast cancer patients.

Guidelines given by health professionals regarding medical intervention side effects prevention were given to only 8% of breast cancer patients. This has severe effects on quality of life of breast cancer patients. It is safe to conclude that all health professionals must be educated about medical intervention side effects prevention and treatment possibilities and options.

According to questionnaire EORTC QLQ-C30 (version 3) result analysis 100% of research patients had satisfactory level of physical scale on different levels, 98% had satisfactory level of daily role on different levels, 92% had satisfactory emotional scale on different levels, 97% had satisfactory cognitive scale on different levels, 96% had social satisfactory scale on different levels while 94% had satisfactory global life quality scale on different levels.

According to the results gained within all the functional scales, the group subject matter of this research has exhibited high values. It is important to note that this research group is specific in type, namely all research patients have completed medical treatment protocol of breast cancer at least one year before joining this study group, they experienced no reoccurrence and are socially active.

70% (n=100) patients experienced varying fatigue symptoms on different levels. It can be said that excessive fatigue can be attributed to habits such as smoking and alcohol consumption as the research group of patients who smoked (n=34) reported excessive fatigue among 73% of researched patients.

This research has shown that within the sample group of patients with limited range of shoulder movement (n= 39) 82% experienced pain on different levels. Based on the symptom scale results, from the total number of researched patients 89% experienced no nausea and vomiting. 68% of patients experienced insomnia on different levels. On the overall basis, 74% patients expressed concern for their future health which badly affects their existing quality of life. 82% experienced no loss of appetite, while 85% had no constipation and 86% experienced no diarrhea. 88% experienced financial burden caused by the illness suffered. It is also important to note that according to the result analysis based on EORTC QLQ-BR23 questionnaire, 88% of patients expressed dissatisfaction with their body image, while 62% experienced no sexual desire.

5. Conclusion

Results of quality of life questionnaire EORTC QLQ-C30 analysis indicate that functional scale analysis results and overall life health status / quality are valid and applicable estimation of our patients' life quality. Similar results were gained from fatigue and pain symptom scale analysis and items estimated were dyspnea, insomnia, and financial impact of disease. Results gained from nausea and vomiting symptom scale analysis and items which estimate loss of appetite, constipation, and diarrhea while taking into consideration the specificity of our group were not relevant. Results obtained from Module QLQ-BR23 analysis indicate that scales of body image, sexual function, estimation of sexual pleasure, future perspectives estimation, breast and arm symptoms scales are valid, applicable, and relevant indicator. Systematic therapeutic side effect scale result as the item which estimates disturbance caused by loss of hair, while taking into consideration the specificity of our group was not relevant. Average life span of researched patients is 58 years. 71% underwent mastectomy with axillary dissection while 79% underwent radiotherapy protocol, which causes frequent manifestation of medical treatment effects, and significantly affects quality of life.

Author details

Sabina Saric
Public Health Institution of Canton Sarajevo, Bosnia and Herzegovina

6. References

[1] Zavod za javno zdravstvo FBiH. Incidenca raka u Federaciji Bosne i Hercegovine 2007. Registar raka za Federaciju Bosne i Hercegovine. Sarajevo; 2008. st. 5-91.

[2] Institut za zdravlje Republike Srpske. Zdravstveno stanje stanovništva Republike Srpske 2007. godine. Banja Luka; 2008. st. 5-14.

[3] Glodner B, Granić M, Kanjuh Ž: Limfedem ruke posle lečenja raka dojke. Beograd; 2005. st. 70-114.

[4] Fayers PM, Machin D. Quality of Life: Assessment, Analysis and Interpretation. J Wiley & Sons Ltd, Chichester. 2000; 12:45-67.

[5] Aaronson NK, Ahmedzai S, Bergman B, Bullinger M, Cull A, Duez NJ, Filiberti A, Flechtner H, Fleishman SB, de Haes JCJM, Kassa S, Klee MC, Osoba D, Razavi D, Roffe PB, Schraub S, Sneeuw KCA, Sullivan M, Takeda F. The European Organization for Research and Treatment of Cancer QLQ-C30 : A quality-of-life instrument foru se in international clinical trials in oncology. Jurnal of the National Cancer Institute 1993; 85:365-376.

[6] Fayers PM, Aaronson NK, Bjordal K, Groenvold M, Curran D, Bottomley A,on behalf of the EORTC Quality of Life Group. The EORTC QLQ-C30 Scoring manual (3rd Edition). European Organization for Research and Treatment of Cancer, Brussels; 2001. p.154-186.

[7] Feuer EJ, Wun L-M, Boring CC, Flanders WD, Timmel MJ, Tong T. The lifetime risk of developing breast cancer. J Natl Cancer Inst 1993; 85:892-898.

[8] Bray F, Sankila R, Ferlay J, and Perkin MD. Estimates of cancer incidence and mortality in Europe in 1995. European Journal of Cancer 2002; 38: 99-166.

[9] Marck P. Lymphedema: patogenesis, prevention and treatment. Cancer Practice 1997;5:32-8.

[10] The World Health Report 2002. Reducing Risks, Promoting Healthy Report. World Health Organization; 2002. p. 8-23.

[11] Speroff L, Glass RH, Kase BG. Clinical gynaecologic endocrinology and infertility. Lippincot Williams&Willkins; 1999. p. 617-28

[12] Drummond MF, O'Brien B, Stoddart GL. Methods for the Economic Evaluation of 12- Health Care Programmes. Oxford University Press, Oxford, 1997; 11:143-150.

Chapter 12

Nutrition Intervention Improves Nutritional Status and Quality of Life Outcomes in Patients Undergoing Radiotherapy

Elisabeth Isenring

Additional information is available at the end of the chapter

http://dx.doi.org/10.5772/55988

1. Introduction

As radiation therapies continue to evolve it is important that supportive care, including effective nutrition support, also improves for best patient care and outcomes. Several sets of evidence-based nutritional management guidelines have been developed for patients with cancer. There is strong evidence to suggest that nutritional counselling by a dietician and/or supplementation is beneficial in improving nutritional status and quality of life in patients with gastrointestinal and head and neck cancer receiving radiotherapy. There is also some evidence to suggest that specialised supplements including omega 3 fatty acids and/or immunonutrition may be beneficial in particular patient groups. In order to provide timely and appropriate nutrition intervention and improve patient outcomes, early and ongoing nutrition screening and assessment needs to be implemented. As new cancer care centres and treatments become available it is important that evidence-based nutritional care is provided as part of multidisciplinary care for best patient outcomes.

2. Malnutrition is common in patients with cancer

Patients with cancer are one of the diagnostic groups at greatest nutritional risk (Watterson et al 2009). A recent observational study in 191 oncology patients receiving cancer services at a public Australian hospital found that almost one half of patients were malnourished, and common symptoms impacting on dietary intake included taste changes, poor appetite and nausea (Isenring et al 2010). Consequences of malnutrition include increased risk of infections, poor wound healing, decreased quality of life and transfer to higher level care (Watterson et al 2009). Malnutrition is particularly of concern as it has been shown to independently lead to increased hospital readmissions and in-hospital mortality, even after adjusting for disease type and severity (Agarwal et al 2012). Strong evidence exists to support the prevention and early detection of malnutrition, with nutrition intervention significantly improving patient and clinical outcomes (Watterson et al 2009).

3. Why can radiotherapy lead to nutritional problems?

While radiotherapy techniques are continually improving they may result in significant side effects to the patient. Radiotherapy has a localised anti-tumour effect, damaging rapidly dividing cells, but can also affect healthy tissue within the treatment field. Radiotherapy acts by directing X-rays to cause damage to cell DNA so cells cannot replicate. Rapidly dividing cells (e.g. blood cells and gut mucosa) are the most susceptible to radiation change. Therefore tumours that require radiotherapy to an area of the head and neck or gastrointestinal tract are likely to lead to nutritional problems. Potential side effects of radiation therapy to the head and neck area may include mucositis, odynophagia, thick saliva, xerostomia, trismus, pharyngeal fibrosis and decreased appetite due to changes in sense of smell and taste (Rademaker et al., 2003). Radiotherapy can also exacerbate tooth decay due to induction of xerostomia and removal of dental floride. Patients with head and neck cancer should see a dentist prior to commencing treatment and decayed teeth should be removed at least 7 days prior to commencement of radiation therapy. Radiotherapy to the thyroid gland in the neck area may lead to hypothyroidism so patients should have their neck area checked regularly (People Living with Cancer 2012). Consuming enough calories to prevent additional weight loss is therefore vital for survivors at risk of unintentional weight loss, such as those who are already malnourished or those who receive anticancer treatments affecting the gastrointestinal tract (Rock et al., 2012). Patients receiving radiotherapy to the gastrointestinal area may experience diarrhoea, constipation, gastric pain, indigestion and/or flatulence which can impact on nutritional status and quality of life.

An aspect of treatment not usually considered is that during radiotherapy patients are required to spend large amounts of time receiving medical treatment and waiting for appointments which can disrupt routines and lead to missed meals. In our experience, rural patients may be at increased nutritional risk as they often need to travel large distances to receive treatment. Their alternate accommodation may not have suitable cooking facilities or equipment such as a blender for softer, pureed foods that may be required if experiencing swallowing difficulties. The patient may not have the energy or skills to prepare suitable foods and fluids during this time. Therefore having an occupational therapist, social worker and/or nurse who can liaise with the patient, care givers, if available, and/or organise home help may be particularly important for the patient at nutritional risk without sufficient support to help with shopping and cooking.

Patients may not be aware that side effects of radiotherapy are often experienced a few weeks after commencement, continue during treatment but may also continue to build and be experienced for 4-6 weeks after completing radiotherapy treatment. This period is an important time for review e.g. a telephone review by a nurse to see how the patient is progressing. Often the patient thinks that the side effects will stop and they will feel better after finishing radiotherapy treatment. However, as the side effects can continue and even become worse for 2 weeks after treatment completion, if the patient does not have adequate support, they can become dehydrated and/or malnourished. These nutritional issues may not be picked up unless the patient is admitted to hospital or when they next come in for a medical review which may be 4-6 weeks later. Therefore a follow up telephone review in the first few weeks following radiotherapy treatment can be useful to identify any problems that may require additional medication and/or support.

Patients receiving radiotherapy to the head and neck area may also experience long term swallowing difficulties. These swallowing difficulties may increase the risk of malnutrition. Therefore ongoing liaison and review by the multidisciplinary team, including a dietician and speech pathologist, may be required.

4. Importance of good nutrition during radiotherapy

The continuum of cancer survivorship raises different nutritional needs and challenges and includes cancer treatment, recovery, living after recovery and for some, living with advanced cancer (Rock et al., 2012). Maintaining nutritional status during anti-cancer treatments including radiotherapy is important for a number of reasons. Significant loss of body weight is not only an indicator of poor prognosis and associated with decreased physical function and quality of life, but weight loss can also affect treatment schedules. Weight loss during radiation therapy to the head and neck can place the safety and effectiveness of the treatment at risk, requiring repeat CT scans in order to keep critical structures to within accepted tolerance doses (Davidson et al., 2006)).

5. Overview of nutrition intervention in patients receiving radiotherapy

There is strong evidence that nutrition intervention improves patient outcomes in patients receiving radiotherapy to the gastrointestinal or head and neck region. Dietary counselling by a dietician and/or oral nutritional supplements are effective methods of nutrition intervention and have been found to improve dietary intake, nutritional status and quality of life in patients receiving radiotherapy (NHMRC grade of recommendation A) (Isenring et al 2012).

6. Head and neck cancer patients

As previously discussed, because of the field of radiotherapy treatment, patients with cancer of the head and neck area are often those at greatest nutritional risk. In 2011, Evidence Based Practice Guidelines for the Nutritional Management of Adult Patients with Head and Neck Cancer (HNC) were released. These HNC guidelines, which cover all treatment modalities (radiotherapy, chemotherapy and surgery), reviewed 288 studies including 45 randomised controlled trials. There is evidence (Grade A) in patients with HNC that supports weekly dietician contact during radiotherapy and at least fortnightly for 6 weeks post treatment, with contact as required for up to 6 months (Grade C).

7. Early identification of nutritionally at risk patients

Medical Nutrition Therapy involves the assessment of nutritional status, nutritional diagnosis and using professional judgement to individually tailor an appropriate nutritional plan. Obviously, the goals and outcomes of nutrition intervention will be dependent on the diagnosis and prognosis of the patient.

Firstly, in order for patients with cancer who are at nutritional risk to be appropriately identified and referred to the dietician, nutrition screening should be routinely used in oncology settings (Isenring 2008). Several valid nutrition screening tools for oncology patients exist, including the Malnutrition Universal Screening Tool (MUST), Malnutrition Screening Tool (MST) and Nutrition Risk Screening 2002 (NRS-2002)(Skipper et al., 2012). Some nutrition screening tools are more detailed than others and designed for different settings and users. Therefore it is recommended to use a valid and reliable tool appropriate to the setting. The simplest tool, the Malnutrition Screening Tool (MST) consists of two questions enquiring about unintentional weight loss and poor dietary intake and can be administered by nursing or administration staff or by the patient themselves (Ferguson et al 1999). In absence of a formal screening system, malnourished patients can be overlooked, especially if they appear normal or overweight (Watterson et al., 2009). Patients identified as at nutritional risk by the MST can then be referred to the dietician or a trained health professional for a comprehensive nutrition assessment.

A valid and reliable nutritional assessment tool for patients with cancer is the scored Patient Generated – Subjective Global Assessment (PG-SGA). The PG-SGA is specific for cancer patients and was developed from the commonly used nutritional assessment tool, Subjective Global Assessment (SGA). Using the PG-SGA tool, patients are categorised as well-nourished (SGA A), moderately or suspected of being malnourished (SGA B), or severely malnourished (SGA C). The PG-SGA also has a scoring system which includes: a patient-completed medical component (weight loss, nutrition impact symptoms, dietary intake and functioning), and a clinician-completed component (diagnosis, age, and metabolic stress). An increase in PG-SGA score reflects greater risk for malnutrition. This system enables clinicians to rank the nutrition risks of individuals within the same SGA category.

Patients deemed to be at very high nutritional risk e.g. head and neck cancer receiving chemo-radiation may bypass screening and proceed directly to nutritional assessment. If a patient is identified as at nutritional risk and there is no dietician available, then it may be appropriate to proceed directly to a nutrition intervention such as high energy and protein diet, oral nutrition supplements or seek further nutritional advice from someone with nutritional expertise e.g. oncologist or physician trained in nutrition. These health care professionals can individualise nutritional recommendations to the patient, but some general tips are described below:

- Small frequent meals
- Make every mouthful count i.e. nutritious foods eaten first, including high energy and protein sources

- Use sauces to moisten meals
- Add fats and sugars to increase energy density and milk/protein powders to increase protein density
- Milkshakes made at home, sauces, commercially available dairy convenience foods
- Commercial high energy and protein supplements, can be dairy, soy or cordial based
- Fortify foods and fluids with protein and carbohydrate powders

If these strategies are not helping to prevent/slow weight loss then more intensive nutritional intervention may be required e.g. enteral (tube) feeding. If the gastrointestinal tract is not functioning appropriately then parenteral nutrition may be warranted. Nutrition support decisions should be made by the multidisciplinary treatment team in conjunction with the patient and carer.

8. Nutritional monitoring of cancer patients receiving radiotherapy

It is important that body weight is regularly monitored and recorded. Even small amounts of unintentional weight loss each week can result in significant unintentional weight loss over a few months. It is easier to slow unintentional weight loss than to try and lead to weight gain after a patient has already lost a significant amount of weight (eg. 5% of body weight in one month or 10% of body weight in 6 months).

Tips for effective body weight monitoring include:

- Measure at the same time of day each week
- Be consistent i.e. before meals, no shoes, in light clothing
- Be aware that different conditions can affect hydration levels influencing body weight e.g. chemotherapy, renal disorders, end stage of disease

It is important that clinicians remember that body weight gives no indication to body composition and fat can mask significant loss of muscle mass. It is this lean tissue that contributes to physical function impacting on patient quality of life. There are many simple bioelectrical impedance devices that give an indication of body composition as well as body weight. Although these devices can be affected by hydration levels, as long as their limitations are understood they can still provide more useful information than body weight alone.

If a health professional trained in nutrition is available then regular assessment of nutritional status by a validated tool e.g. PG-SGA is recommended. Regular nutrition screening, early and timely referral and assessment and intervention by the dietician, as part of the multidisciplinary team and regular monitoring of outcomes offers best nutritional care for patients.

9. Mouth care

Sugar-free chewing gums and sweets, and alcohol-free mouth rinses can help with a dry mouth. Artificial saliva sprays and oral lubricants may be useful, but once again appear to be based on personal preference. Mouth care is important and many centers recommend patients use a made-at-home salt water and/or bicarbonate of soda mouth rinse. For a dry mouth, carrying around a water bottle and sipping frequently as well as keeping a glass of water by the bedside can be beneficial.

• Implement routine nutrition screening e.g. Malnutrition Screening Tool
• Refer high risk patients for nutrition and swallowing assessment e.g. Scored Patient Generated Subjective Global Assessment
• Consider whether patient may require a feeding tube (discuss with the multidisciplinary team, patient, and caregiver)
• Monitor weight regularly (ideally weekly during radiation therapy, every outpatient appointment)
• Aim for weight maintenance (or at the very least minimize weight loss) during treatment
• Manage nutrition-related symptoms as a multidisciplinary team

Table 1. Summary of nutritional management of patients receiving radiotherapy

10. Nutrition and physical activity recommendations for cancer survivors

Cancer survivors are often highly motivated to seek information about food choices, physical activity, and dietary supplements to improve their treatment outcomes, quality of life, and overall survival (Rock et al., 2012). Many patients are interested in nutrition and seek nutritional advice external to the cancer centre. It has been reported that 40% of cancer patients are seeking extra nutrition resources and would like further information regarding dietary tips for managing side effects and supplements (Isenring et al 2010). Therefore it is important that health professionals feel comfortable answering common nutritional queries using an evidence-based approach, have access to appropriate resources e.g. Cancer Council handouts, or can refer to a dietician. The World Cancer Research Report recommends that all cancer survivors receive nutritional care from an appropriately trained professional (physician and/or qualified nutrition professional e.g. dietician) if able to do so. Unless otherwise advised, patients should aim to follow the recommendations for diet, healthy weight and physical activity (WCRF 2010). Patients receiving anticancer treatment can exercise if they wish during treatment but should restrict activity if they are anaemic and should avoid chlorine exposure to irradiated skin (eg, from swimming pools) (Rock et al., 2012).

11. Cancer cachexia

The complex clinical syndrome known as cancer cachexia (from Greek **kakos** (bad) and **-hexia** (condition)) differs from malnutrition in that it is characterised by a negative protein and energy balance, progressive loss of skeletal body mass (sarcopenia), anorexia and metabolic derangements (Dewey et al 2010; Fearon et al 2011). The weight loss seen in patients with cachexia is from both muscle and fat, which is distinct to that seen in patients with malnutrition or anorexia where weight loss is predominantly from fat (Evans et al 2008). This variation is due to the metabolic alterations and inflammatory state that occurs in cachexia (Arends et al., 2006). Cancer cachexia is a multi-factorial syndrome that cannot be fully reversed by conventional nutritional support and leads to progressive functional impairment (Fearon et al., 2011). Cancer cachexia is most commonly exhibited in patients with advanced disease particularly in solid tumours such as pancreatic, lung, gastric and colorectal cancer (Bauer et al., 2005; Dewey et al., 2010). Symptoms may include severe weight loss, anorexia, early satiety, together with associated fatigue and weakness (Bauer et al., 2005; Dewey et al., 2010). Cachexia has a significant impact upon patient morbidity, reduced quality of life and is implicated in 30-50% of all cancer deaths (Palomares et al., 2006).

The nutritional goals and outcomes of patients, particularly those with advanced cancer, need to be realistic, individualised and synonymous with the overall goals for the patient (Bauer et al., 2005). The patient's prognosis and own wishes must be considered with the nutrition intervention adjusted accordingly for those requiring palliative supportive care. The *Evidence Based Practice Guidelines for Nutritional Management of Cancer Cachexia* provides a clear and evidence-based framework to effectively guide nutritional intervention in patients with cachexia (Bauer et al., 2005).

Weight stabilisation is an appropriate nutrition intervention goal for patients with cancer cachexia as it has been shown to improve quality of life and prolong survival compared to patients who lose weight (Andreyev et al 1998, Davidson et al 2004). In order to accomplish weight maintenance in patients with cancer cachexia, it is important to ensure that patients have optimal symptom control and can achieve adequate energy and protein intakes. It has been estimated that an energy intake of at least 120kJ/kg/day and protein intake of approximately 1.4g/kg/day should be prescribed to patients with cancer cachexia in order to maintain weight (Davidson et al., 2004; Bauer et al., 2005). Frequent nutrition counselling (weekly to fortnightly) by a dietitian has shown to improve nutritional and clinical outcomes in cancer patients. The consumption of high protein energy supplements does not appear to negatively impact upon the amount of food consumed (Isenring et al., 2004; Bauer & Capra 2005). In addition, a multidisciplinary approach in order to effectively manage patients with cancer cachexia has shown to be beneficial and further investigation into novel service delivery models is warranted (Glare et al., 2010).

Some studies, though not necessarily nutrition interventions, examined energy expenditure of oncology patients using indirect calorimetry. The Resting Energy Expenditure (REE) ranged from 6300kJ/day (Bosaeus et al 2002) to 8700±1500kJ/day (Cereda et al 2007). In all of these studies, the Harris-Benedict equation significantly underestimated total energy expenditure compared to indirect calorimetry. Therefore there is still a gap in evidence regarding actual total daily energy requirements and at least 120 kJ/kg/day and close monitoring of intake and body weight is recommended. All patients should be considered on an individual basis dependent on clinical factors and discussed amongst the medical team.

Cancer cachexia is challenging to treat. In addition to providing adequate energy and protein intake, other agents have been investigated, including fish oil (eicosapentaenoic acid). Further research regarding the effectiveness of these agents is required.

12. Antioxidants

There are three level I reviews addressing the issue of antioxidant use in radiation therapy and chemotherapy. All reviews (Block et al., 2002; Lawenda et al., 2008; Greenlee et al., 2009) concluded that there was insufficient evidence to provide clear guidelines on use of antioxidant supplements, due to the lack of understanding of the dose-relationship response and the small number of studies on each antioxidant. On the basis of these reviews, it is not recommended that supplemental doses of antioxidants be used, at least until further well-designed studies are published.

As part of the nutritional management of these patients, it is important to not only replenish protein and energy intakes but also vitamin and mineral intakes. Dietary intakes of vitamins and minerals should not be greater than the recommended dietary intakes (RDI) as these may interfere with treatment (Rock et al., 2012). It is important that patients always notify their medical team of any medications and vitamin, mineral and/or herbal supplements they may be taking. Alcohol is an irritant, even in the small amounts found in mouth washes, therefore it is reasonable to recommend that alcohol intake should be avoided or limited in patients with or at risk of mucositis and those receiving radiation therapy to the head and neck area (Rock et al., 2012).

13. Recommendations post radiotherapy

13.1. Long term: Healthy eating strategies for cancer survivors

If there are no long term side effects or nutritional concerns such as swallowing difficulties, reliance on feeding tube or at risk of malnutrition, then general healthy eating and lifestyle recommendations are the same as those for the general population. Key themes include eating plenty of plant-based and whole grain foods, and limiting weight gain (Rock et al., 2012). This may decrease the risk of cancer recurrence (for those with dietary link – over 25% of all cancers including breast, HNC) and other lifestyle diseases such as Type II diabetes mellitus and cardiovascular disease. It is recommended to achieve and maintain a healthy weight, engage in regular physical activity with at least 150 minutes per week and strength training on two days a week and achieve a dietary pattern that is high in vegetables, fruits, and whole grains and reducing energy dense but nutrient poor products such as sugary drinks (Rock et al., 2012).

13.2. Palliative care

Clinicians must always have treatment goals in mind. Therefore whether the patient is receiving radical radiotherapy or non-curative treatment but still may have months or even years to live. In these situations nutrition can still play an important role in terms of preserving lean tissue and body weight and help with activities of daily living and quality of life. If the patient is receiving palliative radiotherapy and is end stage, then the nutritional issues become more about patient preference, eating for enjoyment and quality of life. The use of enteral nutrition and parenteral nutrition support should be individualized with recognition of overall treatment goals (control or palliation) and the associated risks of medical complications and/or ethical dilemmas (Rock et al., 2012). For patients with end-stage disease, focus should be on patient comfort and quality of life (Isenring et al., 2008). Patients with minimal dietary intake may require tube feeding (depending on prognosis and in consultation with patient and medical team) (Isenring et al., 2008).

14. Summary

In conclusion, there is strong evidence to support the benefits of nutrition intervention in improving nutritional status and quality of life in patients receiving radiotherapy. A nutrition screening process should be in place to identify those patients at nutritional risk and put appropriate nutritional assessment and interventions in place. A multi-disciplinary approach is the preferred treatment option as it leads to better patient satisfaction and outcomes. All new cancer treatment centres should include access to an Accredited Practising Dietitian or equivalent (e.g. Registered Dietician) for best patient care. This highlights the importance of early identification and management of nutrition-impact symptoms with adequate follow-up in order to provide optimal care for people with cancer undergoing radiotherapy.

Example meal plans for oncology patients undergoing radiotherapy

Based on a male weighing 70kg. Provides approximately 8500-9000kJ/d (2000-2150Kcal/d) and 110-120 g protein/d.

	Standard High Energy-High Protein (no mouth pain or swallowing difficulties)	**Soft High Energy-High Protein** e.g. sore mouth	**Minced and Moist High Energy-High Protein** e.g. dysphagia, mucositis
Breakfast	1 bowl of cereal with 1 tbsp nuts/seeds and 2 tbsp yogurt, milk*to cover Or 2 eggs (any style) with 2 sl toast, medium spread margarine/butter	1 bowl cereal (without dried fruit or nuts) e.g. porridge, weet bix, rice bubbles) with milk*, Or 2 scrambled or poached eggs on buttered toast (no crusts)	1 bowl cereal well moistened with milk* e.g. weetbix, or porridge or semolina, 2 tbsp yogurt Or 2 scrambled or poached eggs with mashed baked beans
Morning Tea	Cheese and 3 crackers and 1 piece of fruit	Nourishing drink Or dairy dessert	Nourishing drink Or dairy dessert
Lunch	Egg, cheese or meat and salad sandwich (2 sl any bread, includes crusts) I piece of fruit	1 sl. Quiche Or Soft sandwich (e.g. egg and mayonnaise on 2	150g Minced meat and soft cooked vegies (e.g. Bolognaise) or mashed baked beans

		sl. white bread with crusts removed) Soft fruit e.g. ripe banana or tinned fruit	and vegies with extra gravy/white sauce Soft fruit e.g. cut up ripe banana or cubed tinned fruit
Afternoon Tea	1 piece of fruit Nourishing drink	Nourishing drink Or dairy dessert	Nourishing drink Or dairy dessert
Dinner	150 g (cooked weight) Meat or vegetarian alternative and 3-4 vegies Any dessert	200g Fish or soft casserole with sauce Well-cooked vegies Pudding or moist cake with custard/yogurt	200g Nourishing soup or minced/finely chopped stew with extra sauce Mashed banana or stewed fruit (skin removed or finely chopped) with custard/ice cream
Supper	Nourishing drink	Nourishing drink	Nourishing drink

Tips

Milk*Fortified milk=1 heaped tbsp milk powder(FCM or skim) to 1 cup milk

Nourishing drink = 300ml drink made on milk or dairy alternative e.g. soy milk, such as milk shake, hot chocolate or commercial oral nutrition supplement e.g. Sustagen, Ensure. If weight loss is occurring aim for 3 nourishing drinks per day. If weight loss continues refer to nutritional professional e.g. dietician for assessment.

Dairy dessert can include yogurt, pudding, mousse, baked custard, ice-cream etc.

Author details

Elisabeth Isenring
University of Queensland and Princess Alexandra Hospital, Australia

15. References

[1] Agarwal E, Banks M, Ferguson M, Bauer J, Capra S, Isenring E. Nutrition care practices in hospital wards: Results from the Nutrition Care Day Survey 2010. *Clinical Nutrition* 2012; 31 995-1001

[2] Bauer J, Isenring E, Ferguson M. Evidence to support dietary counselling with radiotherapy. *The Journal of Supportive Oncology* 2008; 6 (8); 354-5.

[3] Bauer JD, Ash S, Davidson WL et al. Evidence based practice guidelines for the nutritional management of cancer cachexia. *Nutr Diet* 2005; 63:S3-S32.

[4] Block K, Koch A, Mead Aet al., Impact of antioxidant supplementation on chemotherapeutic toxicity: A systematic review of the evidence from randomized controlled trials. *International Journal of Cancer* 2008; 123: 1227-1239.

[5] Bosaeus I, Daneryd P, Lundholm K. Dietary Intake, Resting Energy Expenditure, Weight Loss and Survival in Cancer Patients. *Journal of Nutrition* 2002; 132(11): 3465S.

[6] Cereda E, Turrini M, Ciapanna D, Marbello L, Pietrobelli A, Corradi E. Assessing energy expenditure in cancer patients: a pilot validation of a new wearable device. *JPEN Journal of Parenteral & Enteral Nutrition* 2007; 31(6): 502-507.

[7] Davidson W, Ash S, Capra S, Bauer J. Weight stabilisation is associated with improved survival duration and quality of life in unresectable pancreatic cancer. *Clin Nutr* 2004;23:239-247.

[8] Davidson W, Isenring E, Brown T, Riddle B. Nutritional management of patients with head and neck cancer: integrating research into practice. *Cancer Forum* 2006; 30: 183-187.

[9] Evans WJ, Morley FR, Argiles J, Bales C, Baracos V, Guttridge D et al. Cachexia: a new definition. *Clin Nutr* 2008;27:793-799.

[10] Fearon K, Strasser F, Anker SD et al. Definition and classification of cancer cachexia: an international consensus. *The Lancet Oncol* 2011;12(5):489-95. doi: 10.1016/S1470-2045(10)70218-7. Epub 2011 Feb 4.

[11] Ferguson ML, Bauer J, Gallagher B, Capra S, Christie DR, Mason BR. Validation of a malnutrition screening tool for patients receiving radiotherapy. *Australias Radiol* 1999;43:325-327.

[12] Glare P, Jongs W, Zafiropoulos B. Establishing a cancer nutrition rehabilitation program (CNRP) for ambulatory patients attending an Australian cancer center. *Support Care Cancer* 2010; 19(4):445-54. doi: 10.1007/s00520-010-0834-9. Epub 2010 Mar 5.

[13] Graham L, Isenring E, Jamieson G. Review of immunonutrition in patients with oesophageal carcinoma. *Diseases of the Esophagus*. 2011; 24 (3); 160-165; DOI: 10.1111/j.1442-2050.2010.01117.x

[14] Greenlee, H., D. Hershman, and J. Jacobson, Use of antioxidant supplements during breast cancer treatment: a comprehensive review. *Breast Cancer Research and Treatment* 2009; 115(3): 437.

[15] Isenring E, Capra S, Bauer J. Nutrition intervention is beneficial in oncology outpatients receiving radiotherapy to the gastrointestinal, head or neck area. *Br J Ca* 2004;91:447-452.

[16] Isenring E, Hill J, Davidson W et al., Evidence-based practice guidelines for the nutritional management of patients receiving radiation therapy *Nutrition and Dietetics* 2008; 65 (Suppl 1): S1-S18.

[17] Isenring E, Cross G, Kellett E, Koczwara B, Daniels L. Nutritional status and information needs of medical oncology patients receiving treatment at an Australian public hospital' *Nutrition and Cancer: an International Journal* 2010; 62:2:220-228.

[18] Isenring E, Zabel R, Bannister M et al., (2012) Updated evidence based practice guidelines for the nutritional management of patients receiving radiation therapy and/or chemotherapy. *Nutrition and Dietetics*. 2013 (DOI: 10.1111/1747-0080.12013).

[19] Marx W, Teleni L, McCarthy AL, Vittetta L, Thomson D, Isenring E. Ginger (*Zingiber officinale*) and chemotherapy-induced nausea and vomiting: a review of the literature. *Nutrition Reviews* 2013(accepted 20/7/12)

[20] Lawenda B, Kelly K, Ladas E et al., Should Supplemental Antioxidant Administration Be Avoided During Chemotherapy and Radiation Therapy? *J. Natl. Cancer Inst* 2008; 100(11): 773-783.

[21] Ottery F, Bender F, Kasenic S. The design and implementation and interventional pathways in oncology. *Nutrition* 1996;12(1 Suppl), S15-9.

[22] Palomares MR, Sayre JW, Shekar KC et al., Gender influence of weight-loss pattern and survivial of nonsmall cell lung carcinoma patients. *Cancer* 1996;78:2119-26.

[23] Rademaker AW, Vonesh EF, Logemann JA, et al., Eating ability in head and neck cancer patients after treatment with chemoradiation: a 12-month follow-up accounting for dropout. *Head Neck*. 2003;25:1034-41

[24] Ravasco P, Monteiro-Grillo I, Marques Vidal P, Camilo M. Impact of nutrition on outcome: a prospective randomized controlled trial in patients with head and neck cancer undergoing radiotherapy. *Head and Neck* 2005; 27(8):659-68.

[25] Rock C, Doyle C, Demark-Wahnefried D et al., Nutrition and physical activity guidelines for cancer survivors. *CA Cancer J Clin* 2012;62:242-274.

[26] Skipper A, Ferguson M, Thompson K, Castellanos V, Porcari J. *JPEN J Parenter Enteral Nutr*. 2012 May;36(3):292-8. doi: 10.1177/0148607111414023. Epub 2011 Nov 1.

[27] Stratton RJ, Green CJ, Elia M. Disease-related malnutrition: an evidence-based approach to treatment. Wallingford: CABI Publishing 2003.

[28] World Cancer Research Fund / American Institute for Cancer Research. Food, Nutrition, Physical Activity, and the Prevention of Cancer: a Global Perspective. Washington DC: AICR, 2007

[29] Watterson C, Fraser A, Banks M et al., Evidence based practice guidelines for the nutritional management of malnutrition in adult patients across the continuum of care. *Nutrition and Dietetics 2009*; 66 (Suppl 3): S1-S34.

[30] People Living With Cancer. www.plwc.org 15 Jul 12.